The Ordering of Time

Plate 1 *'Medieval night'. From a book of psalms, Paris, c.1220, now in the Bibliothèque de l'Arsenal, Paris. Centre: taking the bearings of a star using an astrolabe and telescope. Right: looking up the sun's current position in an Arabic planetary table. Left: the height of the star and the position of the sun are entered in a Latin book; afterwards, these figures are used to measure the hour on the astrolabe.*

The Ordering of Time

From the Ancient Computus to the Modern
Computer

ARNO BORST

Translated from the German
by Andrew Winnard

The University of Chicago Press

ARNO BORST is emeritus professor of medieval history at the University of Constance

The University of Chicago Press, Chicago 60637
Polity Press, Cambridge, United Kingdom

Printed in Great Britain

02   01   00   99   98   97   96   95   94   93   1   2   3   4   5

ISBN: 0–226–06658–4 (cloth); 0–226–06559–2 (paper)

Library of Congress Cataloging-in-Publication Data

Borst, Arno.
  [Computus. English]
  The ordering of time : from the ancient computus to the modern
computer / Arno Borst : translated from the German by Andrew
Winnard.
     p.   cm.
  Translation of: Computus.
  Includes bibliographical references and index.
  1. Calendar—Europe—History.   2. Calendars—Europe—History.
I. Title.
CE6.B6713   1993
529'.094—dc20   93–27993
                                                              CIP

This book is printed on acid-free paper

Contents

Translator's Acknowledgements

I would like to thank Anthony Williams, Mikola Williams, Henry Maas, the staff of Cambridge University Library, and most of all Shirley Winnard for their advice and assistance.

List of Illustrations

Plate 2 *'Modern day'. From a print, Venice, 1496. Ptolemy and Regiomontanus below an armillary sphere. From the fifteenth century this ancient teaching aid was often used as a sundial, with the vertical axis serving as the gnomon and the centre ring (equinoctialis) as the hour-scale.*

1

The Medieval Calendar and European History

In his book *Time: An Essay* (1984), the sociologist Norbert Elias examines the history of the European calendar, providing illuminating comparisons between different epochs. In antiquity, he suggests, people were able to take the few time-signals they needed from natural phenomena; nature's course, however, was 'not sufficiently regular for human requirements'. The modern age therefore constructed a complex system of time-symbols, made by human hands, and designed to fit into the social context of human relations. In Elias's study it is only the Middle Ages which remain obscure. The Church, 'always slow to break with tradition', perpetuated Julius Caesar's calendar and did nothing to improve it. The people of the Middle Ages did not, he argues, differentiate between objective, natural time and subjective, human time, deriving both instead from the same divine creation.[1] The first question I would like to address is whether the Middle Ages really did remain so static while persisting with this antiquated calendar.

In his book *Die Zeit in der Geschichte* (1989), on the other hand, the sociologist Günter Dux regards the medieval period as the most productive epoch for studying the development of modern time. None the less he leaves out of his account the liturgy of the calendar, dealing solely with the economy of work. In the agrarian early Middle Ages, he argues, monasteries continued to live according to the daily routine of the village, with a restricted and short-term logic of action around

which – as in all naturally developed cultures – elaborate
myths evolved. From the twelfth century, it was the combination of skilled manual work and trade in European towns that
first necessitated a calculating approach to dealing with time
and money. In order to survive in the market, tradesmen had
begun to reckon the amount of time invested in their products.
Henceforth the symbol of abstract global time connecting all
simultaneous events in the universe became the mechanical
clock, the prototype of all machines. The physical transformation, however, did not enter popular consciousness until three
centuries later, when it formed the basis of the modern world-
view.[2] The Middle Ages did not, therefore, constitute a *middle*
epoch at all, and medieval man had a mistaken conception of
his own time. The second question I would like to address,
then, is whether we must actually break down the Middle
Ages into an archaic and religious epoch on the one hand, and
a modern, economic epoch on the other.

Occasionally, historians also look beyond the boundaries
separating epochs. In an essay of 1981, Thomas Nipperdey,
from his distant perspective as a modern historian, reveals a
different side to the question examined by Elias and Dux,
discussing the modernity of the medieval conception of time,
the genuinely religious 'Jewish-Christian idea of history with a
direction, heading towards an objective, a history of salvation'.
According to Nipperdey, 'the essence of European civilization
is the way that human beings temporally exceed and tran-
scend their own world in thoughts and expectations of another
world to come . . . While there is clearly a crucial difference
between the eternity of the Middle Ages and the future of the
modern period, and while this future orientation of human
beings – never quite at home in the here-and-now – exceeds it,
this future orientation nevertheless connects medieval man
with modern man.'[3] The third question I intend to examine,
therefore, is whether people in this period passed so resolutely
from their own time into the future that they put aside
thoughts of a day's work and calendrical time.

Elias, Dux and Nipperdey all examine the history of time-
consciousness from the perspective of our present, and do not
credit medieval man with a strong feeling for his present.
Their assumption is called into question by a fact which is

evident to the medieval historian's familiar perspective. The lecture providing the initial outline for the present study was publicized in the following terms: it was to take place 'am Mittwoch, dem 2. März 1988 um 18 Uhr s.t.' ('on Wednesday, 2 March 1988, at 6 pm').* All the time-signals used in this sentence, both verbal and numerical, originated in the Middle Ages; only the apparently most antiquated phrase, *sine tempore,* is modern student Latin.[4] The fact that medieval terms even today serve as rational designations of time would support Max Weber's thesis as applied by the economic historian David Landes (1983) to the notion of time. According to Landes, Benedictine monasticism in the early Middle Ages formed the basis of the modern European measurement and discipline of time by fixing canonical hours and work times.[5] In the very first phase of the Middle Ages, did people really fit out their present day like a prison for sinners and layabouts? This is the fourth question I intend to examine.

Medieval dates and times were fixed according to a procedure that at both the beginning and the end of the period was called *computus* or *compotus.*[6] In 1960 my mentor Herbert Grundmann described the approach of today's medieval historians to the computus as follows: 'Today only a few specialists are able to understand and examine this scholarly medieval computation that every cleric in the quadrivium had to take pains to learn – not because this would be a pointless mental exercise, but because whenever we casually consult our pocket diary we draw on the results of computations that have been corrected over the intervening years.' Our complacency in these matters is so far advanced that modern-day historians discuss time in the Middle Ages and yet forget about the computus.[7] They withdraw money from their bank accounts and type words on to their computers without noticing that the words *Konto* and *computer* are derived from the word *computus.* Conversely, even computer specialists who are historically and linguistically aware know nothing about the past history of the word that constitutes their motto for the

---

* Translator's note: Lecture timetables at German universities are based on the twenty-four-hour clock. Lectures begin either *sine tempore* (exactly at the stated time) or *cum tempore* (within fifteen minutes of the stated time).

future.⁸ What is it that connects the terms *computus* and *computer*, and consequently how much of the Middle Ages is contained within our present? This is the fifth point I would like to address.

We might summarize all these questions as follows: how did medieval Europeans reckon their time, what did they inherit from antiquity, and how much of it did they bequeath to the modern period? In order to answer these questions I shall cross the boundaries between academic disciplines and trace the history of the computus, both the word and the object.⁹

2

Divine, Human and Natural Time in Greek Antiquity

The computus was older than the word used to describe it. People were never able to pace out time, stride across it or delimit it as they could with space; they always had to observe and represent time by means of symbols which themselves required interpretation and could be variously interpreted. No early civilization, however, thought of depicting time by circles, lines or numbers. Without a grounding in mathematics, no one, when confronted by circles that never became round, lines that constantly became bent, and figures that merged and diverged, thought of time as being cyclical or linear. In the main, what humans perceived as time was an uncanny alternation of opposites. Some recurred in natural events such as day and night, summer and winter; others were unrepeatable features of human fate, such as youth and old age, birth and death.[10]

The internal clock of the human organism is not fully synchronized with the external rhythms of nature. Humans cannot completely suppress biological processes: their need for sleep, controlled by the alternation between day and night; their rhythms of begetting and giving birth associated with the lunar month; the times for sowing and harvesting determined by the seasons. But these elementary periods, to which plants and animals are subject unconditionally, can be anticipated or delayed by humans when the force of personal circumstances or even common goals so demand. Humans can contain and

manipulate time within limits because they are the only creatures that actually perceive time. They remain so rooted in
nature, however, that they are unable to co-ordinate measures
of time at will. They have to try to synchronize the rhythm of
their social lives, already complex in themselves, with the
natural cycles of the earth, the sun, the moon and the stars
which, however, do not conform to one single system of measurement. We thus fail to acquire either clear concepts or round
numbers that otherwise regulate our existence. Time can either
be aligned with perceptible experiences, in which case it will
not be consistent, or else incorporated into a logical system of
thought, in which case it will not be accurate.

Historical communities have always drawn differing conclusions from this tormenting conflict, depending in each case
on how they conceived the position of man in relation to God,
nature and his equals. To some, time appeared to have been
fixed for ever, entrusted to divine providence, and remote
from social arrangements, while others conceived time as
something ultimately unfathomable, and yet accessible to a
pragmatic approach, a tool of human communication. To yet
others, it was merely temporarily concealed, shrouded in and
obscured by ignorance, and susceptible to revelation through
steadfast scholarship. All these possibilities were discussed in
ancient Greek culture. As incommensurable notions increased,
rational methods became necessary in order to impose un-
equivocal geometric and arithmetical symbols on the many
different conceptions of time. Out of this emerged that relationship between time and number that was to shape the history of Europe, a relationship as inextricable as it was strained.

This relationship was instituted in the fifth and fourth
centuries BC in the form of three proposals made for different
reasons and with opposing objectives. The earliest proposal,
put forward by Herodotus of Harlicarnassus in his *Histories*,
was based on the experience of the last generation, which was
shaped by the Persian wars. In recounting the history of these
wars *c*.450, the father of historiography had to look beyond the
horizon of the Greeks. He compared entirely different concepts of time and views of history with each other: the
archaeology of the Babylonians and Egyptians with the youth-
ful curiosity of the Greeks; the democratic constitution of

Greek city-states with the monarchical constitution of the Persian empire. On the one hand, there were the officials who came and went from year to year, and on the other a dynasty extending over generations. In the multifaceted world, mortals had only one thing in common: contemporaneity. History was made whenever they came up against each other in their actions and reactions. Herodotus dated them by their relative simultaneity.

Six years after the death of his father Darius, the Persian Emperor Xerxes began his campaign against Greece, a more violent campaign than any before in history. 'Three months after the crossing of the Hellespont . . . the barbarians entered Attica. Calliades was at that time Archon of Athens. They took possession of the forsaken city.' Modern scholars assign a single date, '480 BC', to the two details about the emperor and the archon, but at the time of Herodotus the Greeks had not yet acquired a similar epoch-year, either for their own past or for a univeral time. The date they assigned to the beginning of their communal Olympic games, if any, was '776', '1580 BC' or even earlier. Herodotus was not even able to articulate that '480' was the epoch-year of his present, and that it was central to his record of the three generations that came before and after. The reason for this was that he did not view his world as evolving in a temporal sequence; instead he perceived it as a fragmentation of the names and personal fates of many officials and rulers, and saw it struggle on in the lives of individual people and whole communities from birth until death, from 'zero to zero'. Apart from the dates of people's lives, he discovered neither a basis for, nor any purpose in, extending time and number.

Herodotus knew that the Babylonians divided their day into twelve hours using a sundial and a shadow-stick (gnomon); he was also aware that the Egyptians were the first people to establish the length of the year, divide it into twelve months, and ascertain the month and day of a person's birth with astrological precision. He was cosmopolitan enough to admit that the Greeks acquired this knowledge from the Middle East. Since the Greeks, more casually than foreign antiquarians, also began to count measures of time, Herodotus was able to relate historical events to one another chrono-

logically and hence create 'the cosmos of history out of the
chaos of old wives' tales'. But the Greeks did not recognize any
universal rule for when the year, the month, and the hour
began and ended, contenting themselves with a proliferation
of local customs.

Herodotus bequeathed this pragmatic standpoint to his successors and it has lasted almost to the present day. Statesmen
and generals thought in terms of short-range periods of time
that could be enumerated easily because it was exclusively
among their contemporaries that they found their supporters
and adversaries; the coincidence of simultaneity determined
the extent of their success and posthumous fame. Historians,
too, evaluated political and military events within a narrow
time-frame, and related them to each other using simple numbers in accordance with a particular formula: 'Three days after
one person had done this here, that happened to a different
person there.' Ultimately, neither history nor politics touched
on the combination of time and number. The task of reflecting
upon it was that of philosophy.[11]

After 360 BC, Plato of Athens set about this task in such an
ambitious and thorough way that his *Timaeus* dialogue immediately became the standard work for all theories of time. He
begins derisively by dismissing Herodotus's historico-political
interpretation and recounts the story of the Athenian states-
man, Solon. (Solon had adopted the duration of a human life as
the basic measure of time, divided into ten phases of seven
years each.) According to Plato's account, Solon recited the
local myths of the earliest peoples to an Egyptian priest, and
'by reckoning up the generations, attempted to calculate how
long ago the events in question had taken place (*tous chronous
arithmein*)'. The old Egyptian marvelled at the children's fairy
tales told by the Greeks and produced evidence from his tem-
ple records to show that Solon's ancestors had founded an
almost perfect state in Athens 8,000 years before, and had
repulsed the attack upon the Mediterranean peoples by
powerful Atlantis. Historical time-reckoning was contingent
upon written records – all well and good. However, what was
at issue was not the seventy years of a human life, nor the eight
thousand years in a city's history, but rather insight into the
structure of time itself. Its source was not everyday experience,

but highly abstract theories that could be communicated only in figurative speech.

The Father of Creation filled the universe with movement and life in the image of the eternal gods and planned to make it resemble its model even more. However, since the nature of life was eternal, he was unable to bestow this attribute fully on the created universe. 'But he determined to make a moving image of eternity (*aion*) and so when he ordered the heavens he made an image of eternity which remains for ever at one. We have given to this image, which moves by the law of number, the name of time. For before the heavens came into being there were no days or nights or months or years, but he devised and brought them into being at the same time that the heavens were put together.' When humans, who were not created until later, by other gods, give time a name, and enumerate its parts, they merely translate the fullness of life into its transitoriness. They see in the universe a succession of signs, the origin and undivided reality of which surpass their powers of under-standing. The true 'is' should be reserved for Being; we live in the 'was' of what has become, and in the 'shall be' of what is to become, in the shadow-zones of past and future. The sun of the perpetual present shines behind our backs.

'So that time might come into being, the sun, the moon and five other stars – the so-called planets – were created to define and preserve the temporal numbers.' Plato had learned how to decipher the symbolism of numbers from Pythagoras. He in-corporated Pythagoras's speculations into astronomy even before astronomers knew how to measure the courses of the stars. It was for this reason that he criticized contemporary astronomy. Of the cosmic 'instruments of time', their inter-locking orbits, velocities and numerical ratios, people merely used the three fastest cycles: one rotation of the fixed star-sphere for night and day, one orbit of the moon for the month, and one orbit of the sun for the year. This made their calendar imperfect, even if they added together the lunar month and the solar year to form lunisolar cycles. Mortals overlooked the fact that the most perfect temporal number of all, the Great Year – the year the calendar came full circle – was reached only when all stars and spheres had completed their orbits, after an infi-nite number of years. This world-year was not suitable for the rapid addition of year-numbers, precisely because the inner substance of time was approximate to the image of eternity. Cosmic symbols refer to our eternal ideas, not to our dismal everyday life. 'The sight of day and night, the cycles of months and years, the equinoxes and solstices, has brought about the discovery of number, given us the notion of time, made us inquire into the nature of the universe', and has granted us access to philosophy, the greatest gift from the gods to the human race.[12]

In the public life of Greek cities, knowledge of numbers and time was valued highly for other reasons: it was in various ways useful. People used arithmetic, geometry and astronomy to gain an advantage, in trade as in war. 'A keen eye for the temporal limits of months and years comes in useful in farming and shipping, and no less in the art of warfare', was the opinion of one adventurer. Plato did not begrudge him the pleasure he derived from transitory phenomena, but considered that these sciences diverted the attention of the perceptive man away from what was coming into being and what was passing, towards what 'is' and towards pure knowledge. They were modelled on the divine Creator and His eternal movement; their instrument was the mathematics of ideal symbols and figures.[13] Plato's imposition of doubly inaccessible – religious and mathematical – symbols upon time created a model that would continue to inspire religious and scholarly elites alike until modern times in Europe.

But did these whims of healthy common sense have to remain no more than superficial? They could have at least made the world clear and comprehensible to its inhabitants. If the cosmos was structured hierarchically from the outermost fixed star to the earth at the centre, if its spheres apparently obeyed different laws, then philosophy should not simply talk about incomprehensible and all-embracing phenomena using lofty imagery, but also analyse the principles of sensory perception on earth, and explore the place of time and number in separate areas of human existence.

About 330 BC, Plato's greatest pupil and adversary, Aristotle of Stagira, set out to do precisely that, in a way that would have profound consequences for later epochs, particularly the Middle Ages. Unlike Plato, he recognized separate laws in man's system of signs for thought and speech, as well as a present time, an 'is'. His essay on theorems and judgements divided time along lines suggested to him by the Greek language, beginning with the determinable present of the speaker, resting upon the basis of a definite past, and peering out into an indefinite future. Aristotle did not assign numbers to this philological and psychological time.[14] He dealt with political and historical time in a similar way. His work on politics distinguished two principal phases in constitutional history: the time 'before', when the Greeks still lived under kings and in small oligarchies, and the time 'after', when constitutional states were established and contemporary democracy gained acceptance as the size of the population grew.

Like state structures, events could be compared to one another – or, to be more precise, related to the present – and be viewed from a particular perspective, our own. 'The Trojan War is earlier than the Persian War because it is further away from the present.' In general terms we might say 'that the people of Troy were alive earlier than us, and their ancestors earlier than them', and so on. The lack of an overview of their successive generations, however, condemned them to death: 'Humans perish because they are unable to connect the beginning to the end.' Viewed from a universal perspective, the lives of the dead would be no different from our own, and would constitute a circle without either a before or an after.

Since the philosopher was concerned with general questions, Aristotle criticized the historiographers, especially Herodotus, precisely because they depicted only the particular. Whilst poets such as Homer related the Trojan War as a possibly coherent series of events, a plot with a beginning, a middle and an end, historians merely described it as an actual period in time during which many human destinies ran concurrently with, but independently from, one another. It did not occur to Aristotle to enumerate this accumulation of coincidences using calendar years or Olympiad cycles.[15]

Physical time was another matter. In the *Physics*, Aristotle wrote: 'Time is the number of motion in respect of "before" and "after".'[16] He discussed the common denominator for observed nature and for human perception in his treatise on basic forms of category. Time and number could both be categorized

as quantities, and were arranged according to prior and poste-
rior time: 'in that one part of a time is before and another after'
and in the case of numbers 'in that one is counted before two,
and two before three'. Aristotle was particularly fond of ex-
pressing the various relationships between humans and time
in the image of a man who builds a house. In the act of constructing, he becomes a master builder. He uses the available
raw material systematically, so that it assumes substance and
permanence and so that time is gained, and compensates for
the time he lacks, as his 'inventor and collaborator'.[17]

And what of the stars? Aristotle agreed with Plato's theory
that the sun, moon and stars were in endless and regular orbit,
but he merely concluded that such movements were the
simplest for people to understand. 'We measure movement
by the simplest and fastest movement. For this reason, in
astronomy, we take the most regular and fastest movement –
that of the heavens – as the basis, and judge the others by it.'
We should bear in mind, however, that the heavens provide
people with the shortest unit of measurement – the moment –
rather than the longest, the Great Year. Thus, even when
examined in detail, the laws of astronomy are anything but
simple, and pose numerous puzzles for human sensory
perception. The movement of the sun is regular but the
shadows cast by it on earth firstly lengthen, then become
shorter; at midday, when it is at its highest position, it casts the
shortest shadow. The moon is spherical, but a half-moon
appears as if it were cut off along a straight line; although the
moon comes closer to the earth than does the sun, the shadows
it casts are longer. Why this is the case (and what passes
between the sun and the sundial) can be explained in terms of
physics, but the capacity of natural numbers, the integers and
units of Greek arithmetic, to order the world does not even
extend to the moon.[18]

The three proposals – Herodotus's historical and political
proposal, Plato's religious and mathematical proposal, and
Aristotle's philological and physical proposal – evolved alongside one another for as long as they evaluated the different
strata of a homogeneous world, customs of men, the laws of
gods, and the laws of nature. Their incommensurability did
not come fully to light until late Hellenistic culture had spread

throughout almost the entire known world. It absorbed highly contradictory political, religious, intellectual and economic impulses from all substrata. These included the two most important systems of natural time-division: on the one hand the Jewish calendar, adapted to the needs of shepherds and based on the changing shape of the moon in its phases; and on the other, the Egyptian calendar, adapted to the world of peasants who were guided by the 'orbit' of the sun through its seasons. The two celestial time-keepers could be seen by people with their own eyes, the one more visible by night than the other by day. Anyone who relied on the moon could measure a short time-span using the four lunar phases, and long periods using the synodic month, a period of approximately $29\frac{1}{2}$ days from one new moon to the next. Anyone taking the sun as his point of reference, however, would find his measurements both too small and too large: on the one hand, the swift passing of the day – shorter at one time and longer at another – from dawn till dusk; on the other, the tropical year, totalling approximately $365\frac{1}{4}$ days from one spring to the next.[19]

In order to divide time in a precise as well as a lasting way, both the day and the night were needed, the interaction of sun and moon. As early as the second millennium BC, Babylonian astronomers were looking for direct connections between the sun and the moon. Fifth-century Persians and Greeks already knew that after approximately nineteen solar years – almost 6,940 days – the new moon returned on the same solar day. But even this lunisolar cycle did not work out as a whole number, and illogical steps had to be taken provisionally to adjust it: the lunar months were divided into those with twenty-nine days, and those with thirty, and a thirteenth lunar month was added to seven of the nineteen solar years. Pythagorean-Platonic arithmetic could not come to terms with these complications, and continued to think in whole numbers and simple proportions; Aristotelian astronomy, on the other hand, probably could.

In the 500 years between Eudoxos of Cnidos and Ptolemy of Alexandria, Aristotelian astronomy made dramatic progress, cornering the monopoly on precise computation of times, and developing instruments for accurately measuring time. It integrated the movement of the celestial spheres to the earth's

Plate 3 *The water-clock of Ctesibius of Alexandria, third century BC. Reconstruction by Daniele Barbaro in his edition of Vitruvius (1566), with float, chain-winch, cog-shaft and a dial (modelled on modern astrolabe dials), but lacking the most important component, the regulating system for ensuring an even flow of water.*

fixed climatic zones, thus allowing for no alternative between cyclical and linear conceptions of time. In the third century BC, the inventor of the highly ingenious water-clock, Ctesibius of Alexandria, also had to link the path of the sun astronomically and geometrically with levels of shadow; furthermore he had to co-ordinate the up-and-down movement of the floats and weights with the circular movement of the wheels and hands, both physically and technically.[20] The Hellenistic historians, successors of Herodotus, were fully conscious of the fact that time could not be represented by means of either a circle or a straight line.[21]

The ability of laymen in the ancient world to gain an over-

view of space gave them a sense of rootedness and imposed a
set of limits. They did not live within these artificially fixed
rhythms of short and long duration, but rather in the mid-
range cycle of the seven weekdays. A creation of many different Eastern and Western impulses, the seven-day cycle
provided a clear separation of work and leisure. Neither the
lunar month nor the solar year could be divided by the week
without leaving a remainder; nevertheless, it was possible to
delimit it with the positions of all seven major planets. As well
as the sun and the moon, other stars became increasingly visible to the eye, among them Saturn, the ruler of the first day,
the bringer of bad luck whom people preferred to evade by
doing nothing. 'Saturday' thus became the day of rest, just as
the Jewish Sabbath did for completely different reasons. The
influence of the planets on people's everyday lives was perhaps even more profound than that exerted by political machi-
nations.[22] However, the universality of time could neither be
expressed nor converted into a generally applicable method of
time-reckoning and time-measurement until ancient polythe-
ism and polycentrism had been discarded and the unity of all
divine, natural and human orders had been experienced and
affirmed everywhere.

3

Universal Time and Salvation History in Roman Antiquity

The political conditions for harmonizing the world and time were created in Rome, where the priests' duties included keeping a register of annual officials and feast-days, the *Fasti*. They proclaimed the beginning of each month, the *calendae*, according to the moon's position and recorded the most important events of the year in the *annals*, celebrating them by a *saeculum* of approximately one hundred years, the time-span of the longest human life. Without their pious and scholarly efforts, we would not today refer to *festivals*, *calendars*, *annals* and *secular* events. Yet they could not even agree on the beginning of their city's history, the first year of Romulus, though they did attempt to synchronize it with Greek Olympiad cycles.

In 46 BC Gaius Julius Caesar put an end to the clandestine quarrelling of the hierarchs. Following the suggestion of Egyptian experts, he instituted a pure solar calendar and explained its rules to everyone keen to learn them, thus ensuring not only that the scholarly theories were put into social practice, but also that he gained control over time and that the administration of the Roman empire was standardized. In the long term his calendar reform regulated all notions of universal time. Today, when we refer to a 'leap-day' we honour Caesar's memory just as we do when we say 'July'. He had an immediate influence on the Roman ruling classes, who began to regard precise observance of the hour as a sign of education and power. High-ranking Romans installed Hellenistic sundials

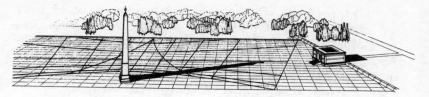

Plate 4 *Augustus's sundial on the Campus Martius in Rome, c.10 BC. Reconstruction by Edmund Buchner, 1976. The height of the obelisk is approximately 30 m. The width of the line-grid, partially excavated in 1980, is over 150 m. On the right is the Altar of Peace.*

and water-clocks in their homes, and had slaves to maintain them. Vitruvius's handbook on architecture showed architects how to construct many different clocks of this kind.[23]

Octavian (Augustus) surpassed his fellow citizens by making the Caesarian tradition sacred. About 10 BC, following the magnificent secular celebration of 17 BC held to mark the dawn of a new age, he erected an Egyptian obelisk on the Campus Martius in Rome to commemorate his recent victory against Egypt and his coming empire of peace. Dedicated to the sun god and the auspicious birth of Octavian, this huge needle of stone formed the gnomon of an enormous sundial, with a Greek line-grid on the floor of the square showing the length of the hours, days and months, together with the signs of the zodiac. On the ground next to this celestial calendar there would probably have been a secular calendar, perhaps with Latin characters: this was Caesar's solar calendar. No one entering the Campus Martius could fail to see that the Caesars united heaven and earth, the Orient and the Western world, and the origin and evolution of time and history, or that they marked the beginning of a universal time.

While the month of August was being dedicated to the deified memory of Octavian, and as more obelisks were being erected as imperial sundials, Romans grew accustomed to dating their politics by the regnal years of the emperor, not merely by alternating consuls. After Livy had used the founding of Rome more than seven hundred years earlier as the starting-point of his pioneering work of history, holding up the city's original greatness to its pampered descendants, historians also reached a consensus on the epoch-year of the empire, *ab urbe condita*, 753 BC. The end of the eighth century was commemorated 'properly' in AD 47; the millennium, on the other hand, was commemorated with less precision in 248, but in the same semi-official manner. Nevertheless, the Julian calendar and the Augustan sun-cult had not yet transformed the working day in the provinces. This would come about only when the Roman world had affirmed its unity religiously as well as conceiving it politically, in the belief in a heavenly power eclipsing all secular misery.[24]

Initially Latin Christians did nothing to foster this process; their empire was not of this world. They regarded the Crucifixion and Resurrection of the Saviour as constituting the focal point of their worship and the starting-point of their sense of time, with the promised return of the risen Christ marking the finishing-point. The task of fixing the date of this beginning using a calendar was made difficult from the outset by the great variety of ancient methods of time-reckoning, and was subsequently hindered by an antipathy towards the Roman cult of the state. Jesus was born at the time of Emperor Augustus (Luke 2:1); he was baptized in the fifteenth regnal year of Tiberius (Luke 3:1) and crucified probably three years later. *Augustus* was also Nero, who saw himself as the centre of the universe rotating by day and night, but during whose reign the apostle Paul suffered persecution (Acts 25:21). The age of the Roman emperors was the mark of their opponents, and at first the Christians believed they would be swiftly overcome (Galatians 4:10). The adoption of universal time was made no simpler by the political victory of Christianity at the dawn of the Constantinian period, in the Roman empire of the fourth century. Three time-systems based on heterogeneous world-views were now to be merged together: the Caesarian order of the solar calendar, beginning at the New Year; the festival of the Mosaic Passover Feast at the first full moon of the spring; and Sunday as the beginning of the week and the day of Christ's Resurrection. The incommensurability between the course of the moon and the the position of the sun, the weekday and the yearly cycle precluded any generally convincing solution. The fixing and implementation of key dates remained what calendrical problems had always been since Caesar, namely issues of power.

Constantine the Great continued along virtually the same lines as the Roman sun-cult when in 321 he designated Sunday as a day of rest and worship, thereby displacing the Roman Saturn-day and the Jewish Sabbath as the beginning of the week. With the unity of the faith, the emperor and the Council also sought to gain acceptance for the standard Easter festival in the Roman world in Nicaea in 325, while overtly opposing Jewish customs, albeit with less determination than posterity supposed. Almost like Plato, the more pious thinkers of early Christendom resisted all attempts to concretize time, and they also had a profound aversion to actual numbers.

To some extent, long-range computation fostered an astrological inquisitiveness that interfered with God's plans, while short-range measurement seemed to delude humans into believing that the moment was at their disposal. Only earthly and peripheral objects could be enumerated and measured, not the lifetime allotted to each human being by God, and still less the period of salvation left to the Church until the return of Christ. Thinkers began to disagree over the Latin word *computus*.[25]

This word did not begin to exert similar fascination upon pagan Romans until late antiquity, when their political control no longer kept the world in check, and when their time was no longer accurately designated by the regnal years of consuls or emperors. They soon acquired the verb *computare*, meaning 'to reckon up', to count on one's fingers, recalling the fact that Roman numerals were modelled on the fingers of human hands. *Computare* accompanied the word *numerare*, meaning 'allocate, count'. Later, a term for using counting beads was added, *calculare* ('to count with numerals'). The noun *numeratio* remained confined to the concept of 'payment in cash'. *Computatio* and subsequently *calculatio*, on the other hand, occupied a wide semantic field, from mathematical 'addition' and economic 'estimation' to social 'assessment' and moral 'evaluation'. Both terms grew to be favoured by Roman lawyers as if tailor-made for a balanced order of communal living.[26] The word *computus*, formed in an analogous way to *numerus*, may have made its first appearance in the third century AD, but it remained superfluous for as long as it meant the same as *computatio*. It was not until the fourth century that it began to denote something different, and then gained wide currency.[27]

The word was introduced with a flourish by Julius Firmicus

Maternus who wrote a textbook on astrology in Sicily *c*.335. 'The same spirit that broke out of the heavenly fire and became involved in earthly weakness in order to lead and guide, handed down this science, these *computos*. It showed us the sun, the moon and the other heavenly bodies that we call wandering stars and the Greeks call planets, their courses and recourses, their positions, conjunctions, their rising and their setting.' *Computus* did not, therefore, have the same broad meaning as the noun *computatio* ('count', 'estimate'), but specifically meant the 'astrological interpretation of computed and observed planetary orbits'. Astrology forged a more solid link between the will of the gods and human destiny than Plato and Aristotle had intended, and showed the reckoner what would befall him and his kind.[28]

This opened up a gulf between pagans and Christians. We need merely contrast pagan astrology with three sentences from the Latin translation of the Bible begun by the Dalmatian Jerome in 383. When the silent sufferer Job cursed the day of his birth, the words he used were: *Non computetur in diebus anni, nec numeretur in mensibus*: 'Let it not be counted among the days of the year, nor reckoned in the cycle of the months' (Job 3:6). This remained an impious wish, for we know that Solomon the Wise said 'that we are accounted Thine', *scimus quoniam apud te sumus computati* (Wisdom 15:2). God accounts to all of us what is ours, and we are accounted his. Anyone who is wise may work out – *computet* – the number 666 of the apocalyptic beast; but this is difficult (Revelation 13:19). Jerome did not himself use the astrologer's term *computus*. More modestly, following Eusebius of Caesarea, his chronicle of 381 numbered the sequence of historical events in years, and traced it back to the creation of the world using biblical dates; *computantur anni . . . a Moyse . . . usque ad Solomonem*. However, Jerome – and in this respect he was Herodotus's pupil – did not locate these historical periods within a continuous world era, unlike some of his Jewish contemporaries who sought security for their homeless faith by ascribing to it a duration of more than four thousand years.[29]

About 400, St. Augustine provided a more fundamental account of the Christian concept of time and numbers, in the best-known chapter of his *Confessions*. He opposed the application of Aristotelian categories to the Creator of heaven and earth. In particular, he dismissed the amalgamation of time and numbers as reification. 'When things past are related truly, it is not the things themselves that are drawn out of the memory, but only such words as were conceived by the images of them which have, as it were, left traces in our senses as they passed by.'[30] The division of language into three tenses met with equal disapproval. 'It may not be properly said that there are three times – past, present and future, but perhaps it might properly be said that there are three times thus: a present time of things past, a present time of things present and a present time of things future.' The weak reflection of a divine omnipresence raised the person of man above animal predecessors. Augustine did not discover the temporal perspectives of 'memory', 'sight' and 'expectation' in the realm of the physical and corporeal. 'For indeed there are in the mind three such times as these, though I do not see them anywhere else.'

Augustine conceived man's relationship to time more artistically than Aristotle, in the image of the singer who produces a song from the soul and projects himself into time. In considering how this sort of time should best be measured, he argued against the oversimplified assumption that 'the movements of the sun, the moon and the planets themselves are time'. If the heavenly bodies served as signs to designate times, they could not be the thing designated itself. Since both psychological and physical time only affected God's creatures, Augustine asked their Creator, 'Do you bid me to agree with him who says that time is the motion of a body? No, you do not bid me.'[31] Firmicus Maternus recommended it. Against this Augustine posited more definite symbols than those of sensory experience or mathematical abstraction – the Word of God – for instance, in a polemic against Manichaeans in 404. 'In the Gospel we do not read that the Lord said: I am sending you the Holy Spirit so that he can teach you about the course of the sun and the moon. He wanted to make Christians, not mathematicians.'[32]

Nevertheless, numbers referred the faithful to the miracle of God's Creation. After 413, referring almost Platonically to the six-day Creation in his work on the *City of God*, Augustine noted that the number six was arithmetically perfect, assembled from the sum of all its parts, its sixths, its thirds and its halves, 1 + 2 + 3. 'It is not in vain that we say in God's praise: Thou hast ordered all things by measure and number and weight.' Augustine took a more sceptical view of terrestrial processes in a world that had forsaken God. When others interpreted the ten persecutions of the Christians in the form of numerical symbols, he called their vain endeavours *computare* and *calculare*, finger-reckoning. For creatures like us, he believed, the historical equation is not solved until after we have passed through the earthly vale of tears. We shall find eternal peace in the hereafter; the six days of the Creation, the six ages of collective wandering, and the six ages in the life of the individual are completed in its number, seven.[33] Until such time, all we can do is identify with God's Law faithfully; we cannot anticipate it by calculation. As Augustine argued in a letter of 419, the date of the world's end is beyond our knowledge. The solar eclipse at Christ's Crucifixion was also a miracle. It occurred shortly before Easter, celebrated by the Jews during a full moon. According to their *computus, astrologi* and *computatores siderum* could not expect a solar eclipse to occur during a full moon. But God was the Lord of time and numbers; hence the word *computus* had blasphemous connotations.[34]

About 500, in the most important Latin textbook on arithmetic, the Christian Boethius wrote with somewhat more restraint, but with a similar, almost Platonic intention, that 'from the beginning all things which have been created may be seen by the nature of things to be formed by reason of numbers. Number was the principal exemplar in the mind of the Creator. From it was derived the multiplicity of the four elements, from it were derived the changes of the seasons, from it the movement of the stars and the cycle of the heavens.' Human sciences, as Aristotle understood them, were separated from their divine origin, and consequently from each other. The science of number dealt with immovable quantities, whereas astronomy dealt with quantities which, instead of progressing in time, revolved in a circle. Neither arithmetic nor astronomy were time-reckoning, and here the word *computus* would be out of place.[35] Like geometry, music is explained by means of ratios between natural numbers; however, while for earth-measurements such numbers merely reveal terrestrial relations, their interpretative capacity raises music to the heavens. Musical sounds and rhythms reproduce the harmony of the planetary spheres and the cycle of the seasons. Even in his textbook on the theory of music, however, Boethius used the verb *computare* only to mean 'reckon', not the special noun *computus*.[36] Had it remained so, the medieval period would never have become the age of the computus.

4

Easter Cycle and Canonical Hours in the Early Middle Ages

Knowledge of numbers helped people to arrange their earthly lives, not towards a possible future, but rather on the basis of the fixed past. In 525 the Scythian abbot Dionysius Exiguus was instructed by the Roman pope to calculate the date of Easter for the following year. This had previously been carrried out by Alexandrian scholars whose Greek writings were translated into Latin. They talked about the *sancte pasche compotum* as if time-reckoning were the arcane knowledge of high priests and specialist scholars, as in Caesar's day.[37] Dionysius dismissed this Hellenistic conceit. He made a clear distinction between the Lord's Easter Day, *dominicum pascha*, and the computed course of the moon, *lunae computus*; the rules for calculating Easter came 'not so much from worldly knowledge, as from inspiration through the Holy Spirit'. The Christian conception of time was influenced only indirectly by natural signs and learned procedures. Dionysius was just as vigorous in tackling the social arrangement of secular dates. He condemned the political custom of dating calendar years by the regnal periods of the Roman emperors, especially after the infamous persecutor of Christians, Diocletian. Instead, Dionysius based his Easter table on the date of the Incarnation: *ab incarnatione domini nostri Jesu Christi*. The perennial feast days commemorated the Incarnation of Our Lord, the occasion of our redemption, and the source of our hope.[38]

Although Christ was the Lord of all time, Christians were

Plate 5 *The Easter Cycle of Dionysius Exiguus, Ravenna, sixth century, now in the Archbishop's Museum, Ravenna. Five nineteen-year cycles (CY. I–V) from 532 to 626, the calendrical dates of the middle of the spring lunar month (L. XIIII) and Easter Sunday (PAS) as well as figures for the position of the moon at Easter Sunday (LU. XV–XXI, a tailed C for VI); at the centre are characters denoting the common year (CM) or lunar leap year (EB).*

able to integrate his unique earthly existence into their own time continuum. Dionysius not only calculated the Easter Sundays for five nineteen-year lunar cycles in advance, from 532 to 626; using other rules of thumb he also related the main Christian feast back to Christ's birth 525 years before, to the oriental lunar cycle with its indicators and the Roman solar year with its leap-days. Ancient knowledge of long spans of time was in this way conveniently presented to the Western world, but it also rendered further inquiry unnecessary. From now on, to find out the dates of the ecclesiastical calendar one had only to look them up in a table; it was no longer necessary to calculate them in advance every year. Why was this? The central point was not the arduous and breathless working day, but the celebration of the Resurrection, the feast without end, when heaven and earth, nature and history were joyously united.

Since the faithful were less intent on exploiting the brief moment spent among their fellow human beings than on re-uniting their immortal souls with their Creator, they found it easy to disregard the hour that had been carefully measured by sundials and water-clocks in the secular Roman empire since Vitruvius. The fact that the hour acquired a crucial sig-nificance for, of all people, the Christians, is due to the Italian abbot Benedict of Nursia, his Rule of the Order written *c.*540, and his table of hours. The abbot's command would on its own have summoned the monks, who lived constantly together, to prayer and work; but Benedict refused to leave the work schedule to the whim of the abbot. As a section of the Rule had to be read aloud to the monastery daily, everyone was able to check the sundial or water-clock to see that their worship was taking place 'regularly'. They were supposed not only to ob-serve a strict temporal discipline in the monastery, but also require it; nothing like it had ever happened before.

Benedict, of course, awarded prominence to the main feasts of the ecclesiastical year, but he omitted none of the working days either. For the daily offices he chose three key points in the late Roman day: the publicly announced changing of the guard in the army at the third hour of the morning (*tertia hora*), at the sixth of midday (*sexta hora*), and at the ninth of the after-noon (*nona hora*). In addition, there were four canonical hours that could be identifed without the need to announce them: sunrise (*prima hora*) and sunset (*vespera*), the dawning sky (*matutina*), and the onset of complete darkness (*completorium*). The psalms were also exactly prescribed, and were to be re-cited every day of the week at these seven times so that the length of each canonical hour could be estimated in advance.

Benedict was just as precise in fixing the hours of waking, eating, working, and resting, and staggered them according to the seasons of the solar year:

During the winter, that is from 1 November (*a Kalendis Novembribus*) till Easter, the time of rising will be the eighth hour of the night, according to the usual reckoning. From Easter till 1 October (*usque Kalendas Octubres*) the brethren should set out in the morning and work at whatever is necessary from the first hour till about the fourth. From the fourth hour until the Sext they should be engaged in reading. After the sixth hour, and when they have had their meal, they may rest on their beds in complete silence . . . The None prayers should be said rather early, at about the middle of the eighth hour, and then they should work again at their tasks until Vespers.

To what end, then, was this order set up? It did not, after all, bless the monks with an ecstatic abundance of time, but simply imposed Rome's civil calendar on them. Benedict's watchword was not mastery of the world, but will-power. 'Idleness is an enemy of the soul'; manual work was merely meant to ward off the gloom of inactivity and in no way enrich the monastery. 'In order that amends be made for sins, the days of our lives are prolonged to give us time in which to make our peace'; collective subordination to the discipline of the Rule elevated the monks above their all too human frailties while leaving them unencumbered by superhuman mortification. Oriental ascetics had not included a midday break after the prayer at the sext – in its haste, Europe has long since forgotten the origins of the *siesta*, together with the purpose it serves.[39]

Did human rules such as these not also help laymen in their profane everyday lives? The celebration of the present sounded more cheerful in the words used by Cassiodorus, the surviving friend of Boethius and Dionysius, to glorify arithmetic as a fundamental discipline *c*.550, with almost an Hellenistic belief in science.

It is given to us to live for the most part under the guidance of this discipline. If we learn the hours by it, if we calculate the courses of the moon, if we take note of the time lapsed in the recurring year, we will be taught by numbers and preserved from confusion. Remove the *compotus* from the world, and everything is given over to blind ignorance. It is impossible to distinguish from other living creatures anyone who does not understand how to quantify, *qui calculi non intelligit quantitatem.*

Astronomy merely succeeded in taking the first banal steps towards measuring time. It helped people 'to conceive the length of the hours, and to take note of the moon's course in fixing the date of Easter'; at best it helped them to judge the weather in different seasons, and to construct sundials, *horologia*, correctly in accordance with each climatic zone.[40] The *compotus*, however, became the symbol for educated circumspection in the midst of barbaric bewilderment.

Nevertheless, clocks continued to be held in high esteem, not as the achievements of Greek exploration or as signs of rank among the ruling Romans, but as evidence of God's miracle of numbers, and as devices to help God's servants plan their time. Throughout the entire early Middle Ages people would read in awe what Cassiodorus wrote to his monks:

> We do not want to leave you in ignorance of hour-measurements (*horarum moduli*); they were, as you know, invented for the great benefit of humanity. For this reason I had two clocks made for you, a sundial (*horologium*) fed by sunlight, and a water-clock (*aquatile*) giving the number of hours constantly, by day and night. For on some days the sun only shines occasionally; then it is possible for the water to do on earth what the sun's brightness, directed as it is from above, cannot do. Human ingenuity has thus caused things separated by nature to act in harmony with each other. The two clocks work as regularly and accurately as if their dials had been attuned to each other. They are provided in order to summon the soldiers of Christ to worship with the most unmistakable of signs, as with resounding trumpets.

Unlike Vitruvius, Cassiodorus did not teach his pupils how to construct these sundials and water-clocks, and consequently they quickly forgot how to keep them going properly. Christian monks were not, of course, supposed to tinker with timekeeping devices; they were expected to study scholarly books, because the duties they performed were far loftier than those of trumpeters of a legion or the clockmakers of a community.

On an earlier occasion, in a letter to Boethius in 507, Cassiodorus had described the method of measuring time by the *horologium* – the sundial during the day and the water-clock at night – as the highest achievement of civilization because the barbarians were in awe of Roman clocks. He now replaced technical skill with arithmetical resourcefulness. Time-reckoning preserved human dignity better than time-measurement.[41] By calculating time, the monks in Cassiodorus's circle at Vivarium were able to learn what no timekeeping devices could teach them: how to humanize daily events and commemorate the history of salvation. In the midst of the decaying Roman world, when Christian laymen sang God's praises and celebrated his supper together every Sunday, and his Resurrection every Easter, they could not just fix upon any day they liked, but had to work out the day God had provided for it.

The group gathered around Cassiodorus produced the first work programmatically entitled 'Computus paschalis', relating to the year 562. Easter Day and time-reckoning were now interlinked, and from this point on, *computus* meant 'the calculation of Easter', referring to both the method of computation and a textbook on the subject. This very early Latin textbook dealt not only with Easter; it adopted Dionysius Exiguus's year-sequence and began by advising: 'If you want to know how many years have passed since the Incarnation of Our Lord Jesus Christ, then reckon (*computa*) thirty-six times fifteen . . .' The product of this multiplication was similarly called *computus*. Anyone who knew the central point – the date of Christ's birth and Resurrection – could determine the time that had lapsed since then by counting on their fingers. Cassiodorus's temporal horizon, however, had become perceptibly more limited. He dispensed with using tables for future Easters and instead updated Dionysius's rules, broadening their application to the working day, weekdays, months and beginnings of the year. Declaring one's belief in the secular present had its price, the sacrifice of intellectual distance and overview.[42]

Pope Gregory I objected to this way of quantifying time in his sermons of 592 and 593. In his view, it was not for arithmetical reasons that the number six was perfect, but purely because God completed the Creation on the sixth day. The speculations of secular wisdom were missing the secret; only the *computus* of someone who uplifted his soul to the Everlasting could grasp it. In matters both great and small it was the allegorical interpretation of numbers, rather than the counting of objects, that pointed the way to heaven.[43] The world ran its course in five world eras according to God's will. Humans went through their lives in four phases from childhood to senility; they divided their day into five segments, from early morning to late evening. This did not, however, produce arithmetical equations. The biblical parable of the vineyard taught us that heavenly rewards are not measured by working hours on earth. Time-reckoning was regarded as foolish because it could not be separated from the mindless measurement of time.[44]

Celts and Teutons, however, newly converted to Christianity, demanded visible signs. They would have especially liked to have the main feast in their faith signalled by a divine miracle. In 577 and 590, when Spanish baptismal fonts filled up by themselves, the miracle preoccupying Latin literature since 444 provided confirmation to the Roman Bishop Gregory of Tours that he had calculated Easter correctly. It was difficult for him to clarify the – for him, embarrassing – *dubietas paschae* by the rules of Dionysius. He was amazed to discover that a slave from southern Gaul was fully versed in the art of reckoning, *ars calculi*. Even for the *subputatio huius mundi*, Gregory had to rely on the Jerome Chronicle; and with his modest amount of arithmetical knowledge he had difficulty in extending the chronology into his own present.

When this first important medieval historian added up the world-years, he arrived at 5,184 from the Creation to Christ's Resurrection, and at 609 years from the Resurrection to the nineteenth year of the king of the Franks, Childebert II. His calculation was even less accurate at the end than it was at the beginning; after all, Gregory did not write these lines 609 years after the first Easter, but 594 years after the birth of Christ. With their short lives, humans obviously could not gain an overview of natural time as a whole; Gregory the historian had enough difficulty in collating everything that happened in his own lifetime from year to year in the vicinity of Tours. When he set out *cunctam annorum congeriem compotare*, from the Creation of the first human being to the present, Herodotus did not so much mean 'compute' as 'relate', and he went on to relate astonishing events in the lives of his contemporaries in a way that was as disjointed as it was thrilling.[45]

Just as overwhelming as liturgical time was natural time, more amazing than all the Seven Wonders of the World worshipped by the ancients (since Herodotus). Gregory wrote his own short book on the subject of time as a divine miracle (c.580). He first celebrated the daily alternation of ebb and flow, the yearly growth of plants and trees, then the daily sunrise bestowing light and warmth upon the earth, the monthly waxing and waning of the moon, and the steady movement of the stars across the sky, some changing monthly, others remaining regular from year to year. Christians should respond to God's miraculous deeds by praising God, not by pretending to know everything. 'I am not here to teach astrology and futurology; I merely encourage that we are sensible in occupying the time taken for the stars to orbit in God's praise. Anyone wanting to perform this service attentively must know at what hours of the night he is to rise and pray to God.'

The fixing of times at night had previously made it more difficult for secular clergymen to adopt Benedictine canonical hours in their entirety. Times during the day were indicated by Cassiodorus's sundials, provided the sky was not completely clouded over. Gregory merely had to take two different facts from ancient books: first, the fact that there were two measures for the hour. The usual measurement, the one that could be observed directly, lasted a twelfth of the solar day from sunrise to sunset, and hence longer in summer than in winter; this was called the temporal hour. The other measure of time divided the circular motion of the fixed star-spheres during one day and one night into twenty-four equal hours; these were called equinoctial hours, because at higher latitudes they could be observed only twice a year, at the equinoxes, and otherwise had to be computed. This was the first thing Gregory learned. The second was the formula for conversion, based on his location in Tours. Gregory ascertained that if the sun shone each month for between nine and fifteen equinoctial hours in the Gallic climatic zone (further south to be more precise), he could use this to discover the approximate length of a temporal hour by dividing the total into twelve and rounding up the result, and could adapt his sundial accordingly.

But what happened at night? If your fellow monk failed to give you the signal to go to the nocturnal offices, you could rely on the moonlight, but only if you knew how to count in fractions. The Franks needed simpler and more reliable clocks than Cassiodorus's water-clocks whose water-flow constantly had to be adjusted or the narrow pipes would quickly become blocked or frozen. And, finally, none of these linear clocks could be calibrated in any other way than by the cyclical movements of the planets. Gregory observed with great accuracy those occasions in the course of the months when familiar constellations on the horizon would rise and fall. He converted nocturnal hours into temporal hours, and determined how many psalms went into each hour. Gregory made no reference at all to the fact that the length of the solar day was less variable in more southern countries than it was in Tours, and that elsewhere the Plough never came above the horizon. Time appeared as fragmentary to people everywhere as it did in our latitudes; it was and would remain the absolute Wonder of the World.[46]

5

World Eras and Days of Human Life in the Seventh and Eighth Centuries

The end of the migration of peoples in Europe marked the beginning of a phase of calmer construction, and acquainted the Teutons with longer spans of time. Mortals still did not have free disposal of the Everlasting Creator's miracles, what they had were formulae; these had to be learned in the schoolrooms of Roman Christians. Isidore of Seville collated them *c*.630. He repeated Cassiodorus's theories almost verbatim and affirmed them, adding: 'If you remove the number from objects, then everything collapses.' In this way Isidore taught people in the early Middle Ages to live in awe of the *computus*; it embraced both the world's course and the human spirit. He also inculcated in them a disdain for *horologia* and ranked such timekeeping devices alongside common tools such as chains and keys.[47]

Although in Isidore's terms, *computare* could mean simply 'add' or 'multiply', anyone who practised universal time-reckoning raised himself far above the *calculator* who collected single numbers like small stones or letters. And studying the *momentum* – the smallest unit of time – involved analysing the movement of the stars because this movement was measured in moments not used on earth (this is faintly reminiscent of Aristotelian doctrines, but more so of the sentence from Acts 1:7, which states that it is not for man to know the *momenta* set by God). It extended from the moment to the Platonic Great Year when all the planets would return to the same point 'after very many solar years'. Nevertheless, God's reckoning, so far as it was understood, consisted of single-figure numbers, the ones that could be counted on the fingers of a human hand. Lacking Augustine's and Gregory's reservations, Isidore telescoped together God's six days of Creation, the six world eras and the six ages of human life, and arranged historical sequences of dates accordingly. The number denoting perfection and totality was to be 7; like Augustine, Isidore reserved this number for God and demonstrated it in his computation of Easter. Natural cycles were in any case closed systems permitting of no human intervention. The times themselves were called *tempora* after the four seasons in which the extremes of humidity and dryness, warmth and cold found a moderating balance, *communionis temperamentum*.[48] Nevertheless, Isidore transformed numbers symbolizing timelessness into arithmetical formulae for secular history. He adopted the years when the leaders of nations lived and ruled as their basic units.

In Ireland, around the mid-seventh century, Isidore's example prompted an anonymous clergyman from the circle of St Cummianus to write the oldest 'computus' in the country. He brooded over the word, spelling it *conpotus* and consequently failing to note its association with *computare*. Instead his thoughts were of *compos*, 'consisting of parts'. In Latin, *conpos* or *conpotus* meant the same as *numerus* – division by numbers in general – and similar words were present in all world languages, spoken by Hebrews, Egyptians and Greeks. However, the only aspect of this science involving counting, *numeratio*, was its general methodology; its specified objective was to study the courses of the sun and moon in order to fix the date of Easter, a problem solved in different ways by those three nations of the world with holy languages, namely Hebrew, Greek and Latin. Irish scholars were keen to take up the subject because it was complex and controversial.[49]

Following Gregory of Tours' example, the Franks applied it more definitely to the present. The so-called Fredegar who, *c.*660, wrote the history of the world and of his people up to 642, began computing the year, *supputatio*, without studying the work of Jerome and Gregory. He did not himself sing the praises of the *computus*, although he did make use of Isidore's work. Instead, using his semi-Roman Latin, he deliberately replaced the verb *comparare* with *conpotare*, where Jerome's chronicle had compared the heroic deeds of the biblical Samson with those of Hercules the ancient. What 'counted' and warranted 'recounting' were the vigorous deeds of men, not the fluctuating processes of time.[50]

However, even they could be made beneficial for present-day purposes. In 678, in awkward Latin, a clergyman traced back the third regnal year of the Merovingian king Theuderic III to Paradise at the beginning of the history of the world, using the dates of years given in the Jerome Chronicle and pompously called *compotum annorum ab inicio mundi*.[51] A short while later, in 727, a Merovingian scholar was persuaded by the Irish 'computus' to derive the entire Latin vocabulary for time-fixing from the primary ancient languages. Unlike the Irishman, however, he did not equate *conpotus* with *numerus*, but with the Greek *ciclus* and the supposedly Macedonian *calculus*. *Conpotus* thus meant the computation of circular motions of all kinds. In spite of Augustine's warning, it was then possible to calculate – *conputare* – the age of the world as a whole, again using Jerome's dating. The addition was clumsily carried out and resulted in 5,928 years. Its continuation, relating to biblical numbers and their mystical arithmetic, went more simply. God had made the world in six days (Exodus 20:11). Before God, however, one day is as a thousand years, and a thousand years as one day (2 Peter 3:8). The world era between Christ and the world's end would also last a thousand years (Revelation 20:7). The consequence was that the world would exist for six world eras, totalling six thousand years. For the Merovingian reckoners, therefore, a full seventy-two years were left over. The editor Bruno Krusch commented mockingly: 'Then came the Last Judgement, when you could blow everything you had.'[52] To put it more kindly, the early medieval Franks looked beyond their present merely in order to get by in it for the time being.

Anyone who regarded the end of the world as incalculable similarly fixed upon dates for the interim stage between earth and Paradise: these dates were days rather than years. Since each Christian was promised resurrection and ascension, the liturgy extended the golden chain from the first Easter Day when Christ triumphed over death, through the feast day when a saint overcame life on earth, to the working day when believers prayed for a late sinner. It was especially imperative that Easter should be celebrated at the correct point in time everywhere, and not with Irish people observing it at a different time from Anglo-Saxons. For holy remembrance days to be observed in a meaningful way, human destinies had to be dated accurately by the course of God's year.[53]

The founder of medieval computation, the Anglo-Saxon monk Bede, was inspired by this requirement. What methods would best meet it, empirical or rational? He had a better understanding of astronomical time-measurement using sundials than Gregory of Tours. About 730 he set out to prove to a contemporary of his that the spring equinox – the earliest Easter date – fell on 22 March and not, as others maintained, on 25 March. He saw the desired result confirmed by *horologica inspectio*, the observation of a sundial whose gnomon cast short and long lines that shifted across a scale. It was also used before Bede's time in England, for fixing canonical hours; several have survived until today. Bede also used a sundial to show that 182 days later, on 19 September, a second equinox occurred. However, he did not reveal details of how he carried out the measurement to his friend.

Elsewhere he seemed to have acquired accurate information from ancient sources about how, in the event of an equinox, the shadow-length of the sundial could be directly converted into the observer's geographical latitude, especially in Britain, but elsewhere too. Bede proclaimed the fact that he kept his sundial under observation throughout the whole year when providing his explanation for the leap-day: after 365 days the sun was not above precisely the same *horologii linea* as it had been a year before.[54] Water-clocks that could divide time abstractly and uniformly without the need for observing the movements of the heavens were completely beyond Bede. He merely wanted to convince uneducated people by means of visible signs, preferring to rely on the two strongest arguments put forward by Christian scholarship: the authority of the Fathers and the rationality of reckoning.

His definitive textbook of 725 was entitled *De temporum ratione*, the reckoning of time. The first chapter was headed: *De computo vel loquela digitorum*. Bede at this time was more

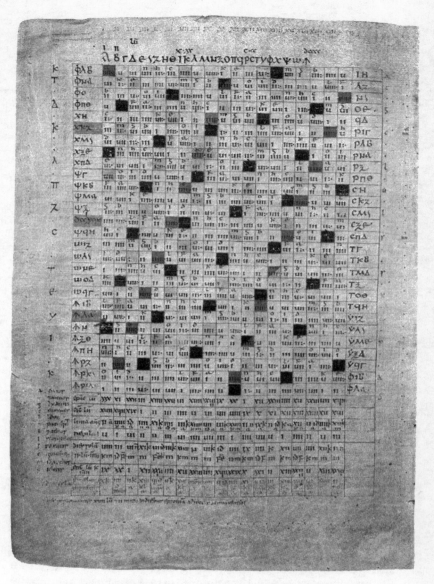

Plate 6 *Bede's calendar-table, Irish manuscript, Laon or Soissons, c.850, now in the Landesbibliothek, Karlsruhe. Each line contains the nineteen years of one lunar cycle; here the number of the weekday on 24 March; above is the moon-letter for determining Easter, signs of the zodiac (in Greek, in the left margin), leap years (with three dots), the first years of the twenty-eight year solar cycle and the fifteen-year indiction (both in shaded boxes) for the entire second cycle from 532 to 1063 (year-dates on the left in Greek letters; on the right, year-dates for the first cycle from 1 BC to AD 531).*

enthusiastic about counting on his fingers than using devices, for reasons less to do with astrological curiosity than with liturgical conscientiousness.[55] Straightforward counting did not achieve very much; long columns of numbers had to be arranged clearly in tables. Bede consequently obscured Isidore's grading system and gave the *computator* (time-reckoner) the name of *calculator*, and later even *catholicus calculator*. Arithmetic – *arithmetica ecclesiastica* – was needed only for church-related purposes.[56] Since arithmetic was difficult to learn and was not allowed to become an end in itself, Bede devised two tables for readers who were either unable or unwilling to count; they indicated the moon's position not by numbers but by letters, recalling Augustine's refusal to regard symbols as realities. When dealing with arithmetical rules, Bede submitted as a teacher to that philanthropic tendency he resisted as a scholar; in order to facilitate reckoning – *calculandi facilitas* or *facilitas computandi* – he occasionally simplified his formulae for complex processes in nature. Bede doubted whether man was able to fix the shifting movements of the celestial lights fully on to the whole numbers he needed for his calendar. This was especially so for the lunar orbit that was central to the computation of Easter and whose measurement (*mensura*) he declared could not be ascertained with any accuracy. By 325 the Nicene Council seemed to have decided on what was most essential and Bede therefore dispensed with arithmetical study, applying the theory of the 'lunar jump' to even out all deviations from the mean value – all too sweepingly, as posterity noted. But Bede's indifference matched his conviction that time, as it was set by God, eluded human estimation.[57]

He was for this reason relentless in condemning calculations of the future in the manner of the Merovingian *computus* of 727, pointing reverently to God 'who, as the Everlasting, created times whenever he wanted, knows the end of times, and puts an end to the fluctuating processes of time when he wishes'. Nevertheless, there was still room left within divine time and natural time for human time. Bede distinguished three types of chronology. The first followed human or divine *auctoritas*. The Olympiads were set by the ancient Greeks, whereas the day of rest at the end of each week was ordered by God himself. Human *consuetudo* prevailed alongside them; the division of the month into thirty days corresponded neither to the sun's course not that of the moon. Other designations of time followed the *natura*, for instance the solar year with its 365¼ days, offering a glimpse of arithmetical reason, the *ratio* of the Divine Creator.[58]

Nevertheless, divine and natural measures of time also proved to be the most human. Bede warned Christian *calculatores* about heathen *mathematici* who he believed were fragmenting birthdates into atoms and assembling them to form astrological prognoses. They needed clocks – *horologia* – with quarter-hour divisions. Christians, on the other hand, needed no shorter measure of time than the God-given hour. For scholarly purposes Bede recommended that they should adopt the twenty-four equal hours from one sunset to the next. Practitioners (Bede spoke of the great masses, *vulgus*) preferred the twelve unequal hours, each depending on the region and the season, and counted them from the Prime at sunrise, through the Sext at midday, to Vespers at sunset; these were the canonical hours of the church and the times for work in the fields.[59]

Bede also discovered a divine and a natural standard for the long periods in human history. The end of our lives and the duration of the world were concealed from us, but God had made the beginning of both clear to us. The date of the Creation, the origins of the interplay between sun and moon, the beginning of the six world eras and of the human species: all could be determined with absolute accuracy by arithmetical, astronomical and exegetic computation. Bede calculated that this date was 18 March 3952 BC. This date pointed the way for history: time was like an arrow shot from a divine source at a heavenly target, and it was still in flight.[60] In this way historiography was made possible, not as universal history but as the history of salvation. Bede expounded this theory in a chronicle at the end of his textbook; it remained in every sense temporary. At least the end of historical change, the Last Judgement in the seventh age of the world, and eternal life in the eighth, were not imminent. Bede computed the dates of Easter for the whole of the second Great Cycle of 523–1063 AD for the next three hundred years.[61]

History mainly affected the years experienced by Bede's

people and church, their coexistence in the warm and bright indoors, sheltered from the cold and dark outside. Bede allowed his fellow countrymen to call the month of the Christian Resurrection after the pagan goddess Eostre, and after the light of Spring that rose in the East.[62] But when did the true light, that never sank below the horizon, enter the world? In 731, in his *Ecclesiastical History of the English People*, Bede replaced the cosmic world era with a human method of dating based on the Incarnation of Christ. It is because Bede's book became a model of medieval historiography that we do not today describe the current year, in the manner of the ancient Romans, as the 2,746th year after the founding of Rome; nor do we refer to it, as orthodox Byzantines and Russians would have, as the 7,501st year after the Creation. Instead, we describe it as the year 1993 AD.[63]

Bede visualized the complete history of salvation in even more direct terms, in the first work of that genre we refer to as 'historical martyrologies'. He was familiar with lists of saints containing thousands of names, in particular the so-called 'Martyrologium Hieronymianum'. This martyrology did not record when, where or how a martyr had attested his faith and consequently what his death had to do with the living. Bede's martyrology compensated for this deficiency by transforming the day of the martyrs' death into the day of their birth for all eternity. He described it thus: 'A martyrology of the festivals of the holy martyrs in which I have diligently tried to note down all that I could find about them, not only on what day, but also by what sort of combat and under what judge they overcame the world'.[64] To this end, Bede consulted all accessible sources, including his own historical works for the recent past. In so doing he noted that the Lombard king had lately, barely a generation before, had the relics of St Augustine transferred to Pavia. Bede concluded the chronicle in *De temporum ratione* with the same hopeful message: the saints of heaven remain with us.[65]

Bede's martyrology inserted a critical selection of 114 saints' names into the course of the ecclesiastical year, as signposts pointing to the life hereafter, just as the verified dates in the chronicle marked the path from the origins of life on earth to the present day. From this point on people in the Middle Ages began making their working days into saints' days. Modern historical scholarship tends to ignore the crucial fact that Bede brought together time-reckoning, the liturgy and historiography; the one cannot be understood without the other two. The computus, the martyrology and the chronicle constituted the three equally powerful mainstays of that scholarship which flourished in Benedictine monasteries and succeeded in bringing eternity into the present.[66]

6

The Church Bell and Work Time in the Ninth Century

The connection between *computus*, martyrology and chronicle was reinforced in the Carolingian period. During this period people were guided by a new time-signal, the bell. Both the German word for it, *Glocke*, and the object itself derived from the Celtic and came to be used by the Franks. Boniface took both with him to the Continent; in English today the time of day is still described using the word *clock*. The hand-bell signalled the daily canonical hours used by clergymen, while the tower-bell summoned laypeople to celebrate mass. Contemporary scholars knew that dividing up the day by sounding a bell was an innovation. Nevertheless they preferred to trace it back to ancient Italy, deriving the Latin name *campana* from the local province of Campania. They impressed on their brethren the times during the day and the ecclesiastical year when the bells had to sound. Bell-time was from the outset more historical than Creation-time or natural time; it was at once liturgical and rational.[67]

In 789 Charlemagne decreed that all priests had to be as familiar with the *compotus* as they were with psalms, musical notes, melodies and grammar, and had to be provided with accurate textbooks on the subject. He reaffirmed this decree relentlessly for the whole empire, and the bishops applied it in their individual dioceses.[68] The emperor set a good example by learning the *ars conputandi* himself and studying the courses of the stars. He regarded the water-clock (*horologium*) sent to him

Plate 7 *The oldest preserved hand-bell, probably from seventh-century Ireland, now in the Collegiate Church of St Gallen. Sheet iron, height 33 cm; the hanging device and clapper are from the modern period; the decoration is eighteenth-century, with the inscription 'Anno 612 hat diss für sein Gloggen gebraucht der Heilige Gallus in seiner Wohnung zu S. Gallenstein bey Bregenz' ('In AD 612 St Gall used this as a bell in his residence at St Gallenstein, near Bregenz').*

from the Orient in 807 as a mere toy. For him, time was not to be artificially manufactured and measured by mechanical means, but was instead to be solemnly observed in the sky and wisely computed.[69]

Between 797 and 799, Alcuin of York provided Charlemagne with written explanations of *computus* and *calculatio* for the solar and lunar orbits. Although the Anglo-Saxon ventured only hesitantly into 'the stamp-mills of the *calculatores*' and the 'soot kitchens of the *mathematici*', he moved their workshops closer together, thereby freeing mathematics from the stigma of astrology.[70] Charlemagne intended to spread his newly acquired knowledge among the people and, following Bede's example, to replace the Roman names of months with more natural names, commemorating the spring in March and the paschal light in April: *Lentzinmanoth*, *Ostarmanoth*. His son, Louis the Pious, dismissed this barbarian vitality, and rather than forgo the written Latin system of dating, he preferred to adopt Mars, the pagan god of war. For this reason we find German poets writing of *Lenzgefühl* (springtime feeling) and *Osterwonne* (the joy of Easter) whereas accounts give the months as March and April.[71]

Curious trial records from 809 show that even Charlemagne's scholarly plans overtaxed his contemporaries. Experts on ecclesiastical time-reckoning were summoned and questioned; they did not fully understand Bede, or anything else besides. Nevertheless, a special name – *compotiste* – was bestowed on them, almost as a sign of rank; Charlemagne immediately ordered that the knowledge they were to acquire should be collected in a seven-volume encylopaedia. Only parts of this mammoth work were copied, even in the empire's centres of learning. It updated three of Bede's tables: the martyrology with the liturgical and computistical dates of the ecclesiastical year; the tables of the lunar cycle and Easter cycle; and the chronicle of world-years up to the current year, the ninth of Charlemagne's empire, the 4,761st year of Creation. A collection of computistical rules provided instructions on how to use these lists. They were accompanied by texts written by ancient authors, Hyginus's description of the constellations, Pliny's theory of the planetary movements and Macrobius's and Martianus Capella's measurement of the earth, sun and moon. Time-reckoning and time-measurement were to be developed into a natural science.[72]

Rather than a global programme of this kind, the time-reckoners needed simple textbooks for discrete areas. In 820 Hrabanus Maurus, a monastic teacher in Fulda, responded to this need with a 'computus' based on Bede's work, and later a 'martyrology' written between 840 and 854 when he was archbishop of Mainz. What we historians today commonly misuse to date his books was conceived by Hrabanus himself as a celebration of the holy present. As a computist he solemnly announced that the year of writing was the year of the Lord 820, the seventh year of Emperor Louis; he even specified the day: 'I am writing today, 22 July.' Similarly his saints' calendar announced that the remains of St Rufus of Metz were taken to the region of Worms 'at the time of the Emperor Lothar', in other words after Louis's death in 840. In this way contemporary history was being written as a history of salvation, and as such it lent a broad dimension to the spatially limited horizon of the living.

This did not move Hrabanus to deal in astronomically large numbers. He simply equated the Platonic Great Year, the orbit of all the planets, with the 532 years of the Easter cycle, the product of the twenty-eight-year solar cycle and the nineteen-year lunar course. The two main celestial lights themselves determined the passage of all the years, months, weeks and days. By day the movement of the sun, and by night, the motion of the stars, provided the *horoscopus* and *calculator*, the traveller and the seafarer, with the smallest available measurement: the hour. At best, sundials were accurate to the hour. But for the nocturnal constellation would there not be a need for smaller units of measurement than Bede permitted? Without using them to count, or measuring them with water-clocks, Hrabanus introduced the atom as the smallest measure of time, of which 22,560 went into an hour, and as the smallest measure of number, the *scripulus*, one 288th of a unit. The pejorative tone of our word *scruple*, however, tells us that Hrabanus and his 'hair-splitting' failed to gain acceptance among the nobility and peasantry.[73]

About 827 in Fulda, then in Aachen, and finally at his home on the Isle of Reichenau where he lived, Hrabanus's pupil, Walahfrid Strabo, made an intensive study of the encylopaedia of 809. He applied its general guidelines to the liturgical and geographical sphere of his own experience, and after some initial uncertainty converted his master's computistical and hagiographical endeavours into verse, when he, too, was at pains to achieve an accurate dating and localization of days.[74] These were followed by related poems that improved the memory by means of euphonic rhythms. Religious education still encouraged the recitation of texts learned by heart, as opposed to the reading of a book in silence. The monk Wandalbert of Prüm, already a proven hagiographer, wrote a martyrology in 848 in ignorance of Hrabanus's script. It spoke of the path from the creation of the *mundi machina* on 18 March 3952 BC to the founding of Münstereifel in 844, combining it with a systematic history of salvation, and a computistical calendar of the recurring seasons and agricultural labours, the names of months and positions of the sun. Inconceivable time was drawing closer to palpable space.[75]

In 863, Agius of Corvey, similarly author of a legend, composed computistical distichs on an Easter table, and in 864, a more comprehensive collection of hexameters. The first verse read: *compotus hic alfabeto confectus habetur*, and by 'alphabetically arranged', it meant eight tables modelled on Bede's containing non-arithmetical letters. In a dedicational poem placed at the beginning, however, Agius celebrated number as the basic principle of Creation, in the manner of Cassiodorus and Isidore, and knowledge of number as the noblest science, because it distinguished between times, from the years right down to the hours, and so allotted people their tasks. Ecclesiastical time-reckoning and scholarly arithmetic seemed to satisfy laymen's need to have their work time divided up and restricted.[76]

The next step on from this was to link the lifetimes of laymen, their memories, lasting impressions, and stories of exemplary deeds, with ecclesiastical time-reckoning. Gregory of Tours's history of the Franks and Bede's ecclesiastical history of the English had already encouraged people to assign the meaning 'to recount' to the Latin word *computare*. The uneducated people of Europe in particular linked the recounting of stories to the counting of time; the vernacular words *conter, raccontare, erzählen*, and *to tell* testify to the rationality of those ordinary people. If asked about the depopulation of their village or a decline in their cattle herds, they replied not with statistics but by recounting a story.

In Latin literature, the relationship between time-reckoning and the art of narration was closest in the Carolingian annals, recording for posterity what occurred during a particular year within the writer's horizon. No annalist needed computistics for his continuous enumeration of years. Nevertheless, a Regensburg scholar who continued the *Annales Fuldenses*, after relating several outrageous incidents in 884, began the next story with the words: *instanti anno, quo ista conpu-*

tamus, 'in the current year in which we are recounting this'.[77]

For a while, however, there was no further encounter between writing and the vernacular, or between time-reckoning and lived experience. The scholars withdrew into their ivory towers. Ado of Vienne followed Bede, beginning a martyrology in Lyon in the 850s and completing a chronicle by 870 when he was archbishop of Vienne. At this point, however, the Carolingian impetus began to diminish. Ado's martyrology added only one topical notice, the moving of All Saints' Day by Emperor Louis the Pious. For the early Christian period, Ado's chronicle took the form of a martyrology and recounted the history of the martyrs; he was not interested in a computistically exact chronology, but in geographical links within his diocese.[78]

The martyrology that the monk Usuard of Saint-Germain finished writing in 865 was ultimately based on Ado's work; it filled the remaining gaps in the ecclesiastical year, and was more concerned with complementing than with confirming the existing dates. It listed several saints for each day, about 1,200 in total, all in the form of short entries, and consistently gave geographical references, but almost never chronological ones. After his journey through Spain, Usuard included Christians who had been killed by Muslims in Cordoba during the 850s. However, anyone unfamiliar with the year of their deaths from another source was forced to regard them as early Christian martyrs. Usuard sacrificed historical fact in other respects too, and in the case of 1 November he even erased Ado's reference to Louis the Pious. The hagiographical part of Bede's programme was thus not so much completed as terminated.[79]

A similar fatigue spread into computation and chronicling. Certain bishops, especially in the western Empire of the Franks, continued to expect their priests to know the *computus necessarius*. Using letters for Sundays and positions of the moon, and numbers for epacts, concurrents and regulars, they were supposed to be able to determine weekdays, beginnings of months, periods of fasting, Easter dates, and main feast days of the year, but *memoriter* – without the aid of books.[80] Anyone who was asked merely for dates seldom inquired as to their reasons. This was the plight of the authors of new textbooks, such as the monk Helpericus of Auxerre. At the beginning of the tenth century, Helpericus wrote a *Liber de computo* of which frequent copies were later made. In the *ars computi* – he also called it *calculatoria ars* – he proposed some reforms, for instance that any *studiosus* who trusted his own eyes more than letters should be able not only to calculate the times of the rising and setting of the sun, but also observe and measure them. The orbits completed by the moon could not be measured in round, whole numbers; first, they had to be reduced to those fractions of time and number introduced by Hrabanus. Helpericus nevertheless claimed to be producing merely an anthology, consisting of older works, especially Bede's; this was, after all, what his public expected.[81]

When the abbot Regino of Prüm began writing a canonical collection *c.*906, he no longer expected clergymen like himself to study Bede's theoretical principles, but only to know the *compotus minor*, the rules of thumb for the current year. Completed in 908, his chronicle showed that he himself did not know how to continue reckoning. It resumed Ado's efforts to expand the early Christian section into a martyrology. For individual years, which he still enumerated by the regnal years of the Roman emperors, he inserted entire litanies of saints who, according to Ado's martyrology, had lived at approximately the same time. Regino created the impression that the entire time-span of the chronicle since the birth of Christ had been accurately computed, *usque in presentem annum, qui computatur a prefata incarnatione Domini nongentesimus octavus.* He counted so poorly, however, that he fundamentally failed in his attempt to synchronize the cycles after Christ's birth established by Dionysius Exiguus with the regnal years of the Roman emperors and Christian popes. Wandalbert's practical connection between the monastic feast day and the agricultural working day also fell by the wayside.[82]

However, scholars redirected their efforts towards related fields that were more resistant to spatial integration and demarcation. The rise of Gregorian choral music in Carolingian minsters necessitated a clearer definition of musical sense. Since Augustine, music had been associated with time, and since Boethius it had been based on numbers. Bede, on the other hand, dealt with it only in passing. In the 840s the monk

Aurelianus of Réomé wrote a 'musica disciplina', dedicated to one of Charlemagne's grandchildren, the first original text-book on the theory of music since Boethius. As Aurelianus maintained in his teachings, all the laws (*rationes*) of musical art consisted of numbers. Recalling the seventh-century Irish 'computus', Aurelianus claimed that what Latin scholars called 'computus' in fact meant nothing other than *numerus*. Like arithmetic, geometry and astronomy, music fell within the competence of natural science. It could be heard in the spherical harmony of the planets, the proportions of harmonizing sounds, the rhythms of successive sounds, and the structures of the eight church modes first declared by Aurelianus. In consequence, the relationship between church music and time-reckoning grew closer over the next two centuries. Mnemonic psalms, for instance, developed from poems by Walahfrid and Wandalbert which scholars have only recently discovered and deciphered, began to be used for computation.[83]

Liturgical rapture was even exposed to growing rational demands in the announcement of local public holidays. About 896, Notker the Stammerer, a monk at St Gallen and an important musician, began composing a martyrology, conceiving it as an abridged version of a variety of ecclesiastical histories. He used historical and computistical arguments to cast doubt upon numerous dates in the church calendar, for example, the traditional but by no means undisputed anniversary of the death of the evangelist Marcus as observed by the neighbouring Abbey of Reichenau. It was directly dependent on when the earliest Christians in Egypt had celebrated Easter, and that was not an easy date to discover. Notker also moved the dates of more recent feast days, such as that of St Afra of Augsburg whom others had confused with a Mesopotamian martyr called Afer. He advised against the impulsive introduction of new saints' days, such as that planned by Archbishop Hatto of Mainz for the cult of St George at Reichenau.[84] Rationality and actuality became the basis for harmonious worship. In this way the Carolingian *computus* required the monks not only to worship God devotedly, but also to show critical vigilance towards even modest regional requirements, including a new concept of time.

7

Perception of the Moment of Respite in the High Middle Ages

The tenth century, still regarded by some historians as a *saeculum obscurum*, paved the way for the breakthrough of the incipient rationalism of the Carolingian period, but in doing so it brought about a fragmentation of the European concept of time. From this point on, the small number of experts spoke of time and numbers differently from the mass of laymen. The canon law could no longer even demand that every priest had to master the computus. Critical time-reckoning was required more than just the formulae of the *compotus minor* learned by heart; books containing reference tables were needed. An increasing degree of specialist mathematical talent was required and there was less and less need for the general historical education widely shared in the Carolingian period and from which the emerging nations of Europe were already distancing themselves. They placed their faith in images of history whose emotional, experiential and narrative times did not have to be computed, thus provoking the rationality of the scholars still further.[85]

Was it still worth the effort? One number threatening to put a quick end to all time haunted people's hearts more than it did their minds. It was no longer based on the naive tricks of Merovingian computists who had dated the Last Judgement by adding six thousand years to the Creation. The prophecy of the Secret Revelation (Revelation 20:2–10) sounded at once more mysterious, more precise, more Christian and more historical: the Devil was chained up for a thousand years after Christ and could not seduce the nations. However, the Antichrist would then be let loose and would torment the community of saints and restrict the spread of the Faith, and hence provoke God's Last Judgement upon the world. Most theologians since Augustine had warned against a computation of the Day of Judgement because no one knew 'that day and hour' (Matthew 24:36). Rather than falling exactly 1,000 years after the birth of Christ, it was not meant to happen until perhaps a generation later, at the millennium of Christ's Resurrection.

Most computists since Bede had attempted to establish the date of the Creation diachronically, but since Dionysius Exiguus they had taken care to avoid synchronically integrating the first Christmas Eve and the first Easter morning into the time of the Roman empire. They did not want to subject Christ to the earthly history he had overcome. They thus also left open the question of when the Son of Man would return to pronounce judgement over the living and the dead. While it was light, Christians constantly had to live in anticipation of the coming of their Lord and appear before their Eternal Judge immediately after their death. They were always living on borrowed time, independent from the history which permitted some nations to survive for the time being.

Now, however, the constantly alternating rhythms of European daily life, the generations and the seasons, the ecclesiastical year and the canonical hours, were themselves severely disrupted. In the late ninth century the mission in Europe began to falter; for the first time since the end of the migration of peoples, expansion ceased and went into reverse. Non-Christian peoples penetrated the Carolingian empire, Hungarians from the east, Vikings from the north, Arabs from the south and west, harbingers, it seemed, of the Antichrist whose apocalyptic hour struck as the tenth century drew to a close. Signs such as these could not, however, be clearly deciphered. The same Arabs who laid waste to the Christian north of Spain, occupying Barcelona in 985, had both a knowledge of, and instruments for, reckoning and measuring time, and their accuracy was a source of amazement to Latin Europeans. Hopelessly though Islam may have confused its knowledge with astrological illusion, Arabian arithmetic, geometry and astronomy did not merely preserve the legacy of Greek antiquity, beginning with Ptolemy's encyclopaedia, but also the remaining knowledge of salvation brought by the patriarch Abraham from Mesopotamia to Egypt.

Did this divine gift of natural science arrive in the West in time to rouse bewildered Christians? Nowhere could God's plan of salvation be seen more clearly than in the signs of the cosmos he created: the solar year, lunar month and sidereal hour. Anyone knowing how to interpret these signs correctly would have been able to restore the world in accordance with the Creator's guidelines, reform Christ's Church, and transform the impending end of humanity into a new beginning. It was simply a question of perceiving the moment of respite, of exploring it by scholarly means and putting it to pious uses.[86]

In 978 Abbo of Fleury began to criticize Bede's dating of the Crucifixion and the Creation, in a 'computus vulgaris' that was anything but vulgar. Abbo applied his arithmetical competence as a *calculator*, demonstrated in his commentary on a late antique book of arithmetic in tabular form. Following in the footsteps of Plato and his late antique descendants, he attributed more importance to time than Augustine and Bede. Time attracted him as a problem for human reflection, rather than as an arena for human activity. Time was a spiritual form rooted in the unity of God; it was quantitatively set out, but beyond human sensory perception. It could not be grasped physically like five coins, yet it could be counted and divided in exactly the same way, only in a scholarly manner that could have no more than an indirect effect on daily life. Abbo did not use the sundial with its visible yet fluctuating temporal hours to illustrate this; instead, he used the water-clock (*clepsidra*) that could be used to divide time abstractly, but regularly, into equinoctial hours. It provided anyone observing the firmament at leisure with fractions of hours that the common man neither sensed nor needed. Nevertheless, the monotonous flow of its water, together with the rotation of the constellation of fixed stars, offered the surest measurement of the annual alternation of days of sunshine and hours in our lives. In Benedictine monasteries the most important time of all was the beginning of the day, the nocturnal hour of waking, and in Fleury Abbo actually stipulated that a water-clock should be used to keep this time, as Cassiodorus had done in Vivarium. It gave those on duty the signal to ring the hand-bell in the dormitory. Like Cassiodorus, Abbo did not deign to explain how this overflow basin was to be fitted.[87]

Computists and historians tinkered in a far more imprudent and concerted way with much longer spans of time than the philosophers and astronomers as if, like God, they had the unity of all times and numbers at their command. Consequently, their conjectures contained much more serious errors, and Abbo was merciless in exposing them. If 21 March, the day when the founder of the Order, Benedict of Nursia, died, had really been an Easter Saturday, then Bede's tables were out by twenty years. Although Bede had counted the calendar years correctly, he attributed the wrong historical date to them. But anyone wanting to establish the time of the Creation could not rely on the historians, the *historiographi* and the *chronographi*. Abbo made a clear distinction between the reliability of nature, *naturae ordo*, and the credibility of tradition, *historiae fides*. The fact that prevailing opinion mistook human time for natural time led to confusion. Bede's long-standing reputation had to give way to a new truth that was now coming to light.[88]

In order to hand this truth down to posterity, Abbo designed a perpetual calendar consisting of elaborate tables with numbers and letters modelled on Carolingian figurative poems, but these were far too long to be learned by heart. The work begun by Bede and taken up by Agius was made by Abbo into a principle. The *calculator* was able to compute time using various symbols, not only fingers and numbers, but also letters, each representing one day; he used the whole alphabet to divide up the year: *Idem alphabetum tredecies computatur in uno anno*. The arithmetical, geometrical and 'literary' regularity of these symbols radiated beauty; this was not a sensuous but rather a reflected beauty, that of mathematical forms expressed in the unfragmented numbers of whole days, months and years. Abbo's study of late antique commentaries on Plato's *Timaeus* was not in vain.

It was no longer the authority of God that was directly reflected in time and numbers, as it had been for Isidore of Seville, but rather natural order on the one hand, and historical custom on the other. The reformer Abbo wanted to reconcile them anew, and thereby close the gap between contemplative monks and active laymen. It was for their sake that he even accepted the 'foot-sundial' devised for peasants in late Roman antiquity. The 'foot-sundial' used the observer himself as a gnomon. He had to count how many footlengths fitted into his shadow by pacing them out, and was then able to read off the approximate hour of the day from a table. The numbers formerly ascertained for Italy went astray north of the Alps; a lot still had to be done to correct tradition. Undaunted nonetheless, Abbo carried out the most important computation, that of the Easter dates for the entire Great Year beginning soon afterwards, the third Easter cycle from 1064 to 1595. He protested that God alone knew whether the world would complete this third cycle, but he had no doubts that the world would begin it. Whatever confronted the human species, this much was certain: 'Just as the first year of the Incarnate Word and then the 533rd year have been, so too will be the 1,065th year.' [89]

The interlinking of nature and history through arithmetical reason met with resistance from both sides, from forces of rapid progression and those of restraint. In Abbo's lifetime Arabic treatises on the astrolabe, an astronomical instrument, were translated into Latin in Christian northern Spain. In order to introduce the astrolabe into the West, Lupitus of Barcelona wrote a preface c.980 objecting to the Church's prejudice against astrology. He argued that astronomy was the study of the *superna machina*, God's Creation, and that the astrolabe was indispensable to correct worship. 'For every clergyman must learn the *computatio* of past and future times in assiduous contemplation and true study, so that he can state the beginning of Easter and the dates of other feast days correctly, both for himself and for others. Also, so that he can sing choral prayers at both day and night at the proper hours . . .'

The astrolabe was thus meant to improve, if not replace, mnemonic rather than arithmetical methods of Latin time-reckoning, the basic formulae used by Gregory of Tours, beyond which Abbo of Fleury scarcely progressed. The instructions that followed, possibly translated by Lupitus, presented the astrolabe pointedly as a sundial, *horologium*, and employed the verb *computare* wherever the user was supposed

Plate 8 *The astrolabe of Muhammed as-Saffar, Toledo, 1029, now in the Staatsbibliothek, Berlin. The front of the oldest preserved astrolabe from Islamic Spain, brass, diameter just under 8 cm. Calibration on the outer rim, rotating 'spider' (eccentric zodiac) with twenty-nine sidereal signs. The inserted plate is also preserved, with contours (above, for latitudes between the equator and the limit of the inhabited earth at 66 degrees) and (below) the curves of the unequal hours and the five Islamic prayer-times.*

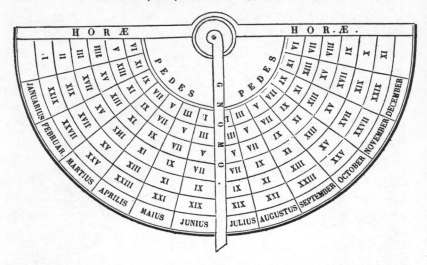

Plate 9 *The* horologium *of Abbo of Fleury, c.1100, redrawn from Abbo's manuscript, now in the Staatsbibliothek, Berlin. Figures for human shadow-lengths in multiples of human foot-lengths at the hours of the day VI–I and (identical with them) VI–XI in the twelve months of the year.*

to count or add in order to determine days and hours. This primarily involved converting the ostensible length of a solar day and its twelve temporal hours familiar to the great masses into the true celestial time of the fixed star-sphere with its twenty-four equinoctial hours imagined by the scholar Bede. The astrolabe not only measured the motion of the sun and fixed stars by means of a sighting device and marked angles, it also worked out the two time-measurements for each position: first, the temporal hours by means of curves in the inserted plate, and secondly the equinoctial hours with straight pointers for calibrating the outer rim. It thus translated the subject of empirical observation quickly and accurately into rationally correct values. This led Lupitus and his followers to turn common linguistic usage on its head, describing the temporal hours pejoratively as 'odd' or 'artificial', while for the equinoctial hours, they used the positive terms 'even' or 'natural'. If the first set of instructions went astray with this reversal, the only remedy was more intensive practice; later texts added the verbs *calculare* and *numerare*, omitting the reference to feast days and canonical hours. Informed work on

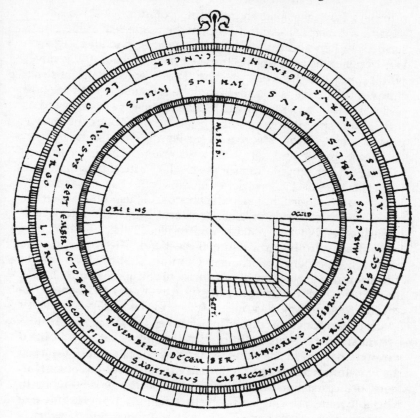

Plate 10 *Back of an astrolabe. Drawing from a manuscript anthology, Fleury, late eleventh century, now in the Vatican Library, Rome. The outer scale is for the signs of the zodiac, with 360 degrees; the eccentric inner scale is for the annual calendar with 365 points; in the centre is a shadow-quadrant for geometrical measurements.*

the new device did more to encourage the extension of a natural-scientific system of symbols than the reorganization of a symbolical system of worship.[90]

Abbo's contemporary, Gerbert of Aurillac, requested and evaluated the treatises from Spain and, following in the footsteps of Aristotle and Boethius, went further in separating the spheres, challenging Abbo's philosophical elevation of the day and the hour. Eternal reason and real necessity were combined in the endless movement of the heavens and the sun; time and numbers, however, had an indefinite potential, and would not assume a concrete form until they were made real and circumscribed, 'when you say day, month, year or suchlike'. Reason was occasionally possible in these time-spans, but it was never wholly and conclusively present. The even and yet abstract times of the heavens could scarcely be reconciled with the odd and yet actual times on earth. Consequently, Gerbert cared as little about speculative cosmology as he did about the principles of computation. His reason for wanting to use arithmetic, geometry, music and astronomy was merely in order to provide an expert explanation for particular natural phenomena.[91]

Gerbert needed instruments for this, not only the astrolabe and the monochord, but also the abacus. He used the abacus as a table, with columns of ones, tens and hundreds into which counting beads were inserted and moved. When Gerbert performed a multiplication using these beads, he modelled the resulting numbers on finger-reckoning, calling them *digiti*, 'finger-numbers', when they represented figures from 1 to 9, and *articuli* 'joint numbers', for figures higher than this. The abacus was the first reckoning device in the European history of science to function 'digitally', in other words, giving results by means of separate numerical symbols; it symbolized shrewd analysis in the midst of crude generalization. It paved the way for an ahistorical rationality that brought about a shift from the consistency of written material to the evidence of readable material, and led to the increasingly important use of fractions, *minutiati*, as well as whole numbers, *integri numeri*.

The term Gerbert used to describe anyone employing this abacus – to check geometrical propositions, for instance – was *abacista*; but an *abacista* was not a *compotista*. Simple time-reckoning by whole numbers, the sort Bede had carried out by finger-reckoning, held no appeal at all for Gerbert. Neither, however, did he consider using fractions that the abacus could have helped to introduce. If he expressed time in numbers, then it was in high numbers. In 996 he wrote elegantly to Otto III: *Extremus numerorum abbaci vestrum definiat*. This, the highest number that could be represented on the abacus, was meant to indicate the length of the emperor's life; it comprised twenty-seven figures, each greater by a power of ten than the one before. It made little difference whether this ideal and inconceivable quantity concerned years, days or hours. Where more definite designations were involved, Gerbert dispensed with both the abacus and the astrolabe, contenting himself with the old basic formulae for temporal hours. In 989, like Abbo, he proposed the use of water-clocks to help draw up lists of hours, *horologia*, and measure the temporal hours. However, he merely noted the full hours, going no further than the solar year. The task of measuring visible objects gradually assumed importance; but the calculation of things perceived by the mind was something different and more pressing.[92] Gerbert bracketed computation outside arithmetic, while his successors adopted it for mathematics.

In a treatise entitled 'On the Use of the Astrolabe', possibly written by a pupil of Gerbert after 989, the verb *computare* was used because, in order to locate fixed stars using this instrument, numbers had to be added together. It made only passing mention of the use of the astrolabe for ecclesiastical purposes and avoided the term *computus*. Nor, when it referred to a *calculator* (reckoner) did it mean a person, as in Abbo's writings, but rather the 'small protruding tip on the outer rim of the spider at the beginning of the astrological sign of Capricorn; when the spider rotates, it travels along the calibration on the outer circle of the "mother plate" and indicates calibrated values there.' The older instruction books sent from Spain had been familiar only with the Arabic name *al-muri* (pointer) spelling it Almeri. The device helped to determine the length of unequal hours and convert them into equal hours without the need for Gerbert's proposed measurement using water-clocks; 360 degrees of the circle = 24 hours; 1 hour = 15 degrees.

This method of gauging time was described as if the pointer itself did the reckoning: *quotcunque partes infra suos limites computaverit*. The new terminology signalled a basic and momentous discovery: if an instrument was properly constructed and set, its analogue dial would relate various infinitely variable quantities to one another and relieve people of the tasks of memorizing and calculation. It did not just reckon for them, it reckoned better than them. The astrolabe became the earliest 'analogue' reckoning device in the European history of science, a modern rival to the computus, and a symbol of concentrated harmony in the midst of feudal dissipation.[93]

Both the astrolabe and the abacus widened the scope of mathematics and reinforced the collaboration between its various specialisms, arithmetic, geometry and astronomy. On the one hand, time-reckoning was still just one of many possible applications; on the other it was not annexed to astronomy, even if the astrolabe was more helpful to time-reckoners than the abacus. This was how the author of an arithmetical treatise from Würzburg conceived it. His treatise took no account of computation and merely touched upon *abaciste*. Nevertheless, *c*.1030, he addressed them as *compotiste*, as though anyone who slaved over an abacus could be so regarded.[94] Franco of Liège wrote a treatise on the squaring of the circle *c*.1046, combining geometrical and arithmetical methods, and self-evidently using the term *computare* to refer to the work carried out by the *calculatores* on the abacus. Computistical terms and methods began to disappear and become subsumed under those of general mathematics.[95]

Objections were raised in conservative monasteries, for example in St Gallen, where *c*.1010, the monk Notker the German wrote a treatise entitled 'On Four Questions regarding the Computus' in order to ward off one pupil's craving for knowledge. The younger man, probably the historian (as he was to become) Ekkehard IV, wanted not only to become a *compotista* but also, like Helpericus of Auxerre before him, a *scrupulosus calculator* by dividing the time-spans of the computus, especially the lunar cycle, into fractions of hours. He felt that the most serious source of error in the Christian calendar was the use of whole numbers. Bede had warned against fragmenting the universe by reckoning; but Notker felt the need to invoke visible signs in order to defend the basic formulae of computation more insistently than Helpericus: after the 'lunar jump', the moon's position in the sky was exactly as Bede had calculated it. Ekkehard was apparently convinced, and became the hagiographer and historian of his abbey without further cultivating his computistical tendencies or, indeed, without reinforcing them by using the astrolabe, an instrument with which he was well familiar.[96]

Nevertheless, even Notker no longer religiously used the computists' specialist terminology. When translating psalms prior to 1020, he discovered the abstract enumeration *quarta sabbati* for the fourth weekday at the beginning of the 93rd psalm. He translated *in mittauuechun* (in the middle of the week) more graphically to indicate the beginning of the week on Sunday and displace the pagan God Mercury. As a consequence of Notker's decision, the Germans call this day of the week Mittwoch (mid-week) rather than 'Mercury Day' as most Europeans do, although to modern ways of thinking which have Monday as the beginning of the week, it is no longer in the middle.[97] Notker's approach was not as mechanical as this: he would not allow a gulf to develop between the theory of reckoning and the practice of observation, between learning and colloquial language.

A Benedictine monk at Reichenau, Hermann the Lame, set out to explore the existing differences and then reconcile them. In his early work 'Musica', written *c*.1030, he was incredibly accurate in identifying the two mainstays of any science: 'the unanimous judgement of all and the insuperable truth of nature'. Both revealed the shared basis of time-reckoning and music theory: the seven notes recurred like the seven days of the week, whose order constantly changed while the days remained the same. The naturally ordered *structura* of all intricately formed life consists of numbers. Hermann, of course, knew that melodies and rhythms appealed 'to passion as well as to reason' and he no longer referred to the 'compotus', as Aurelianus did.[98]

Hermann similarly avoided using the technical term 'compotus' in arithmetical and astronomical studies, although he did use instruments like the astrolabe for measuring time, calling them *horologium*, and adopting the term *calculator* to describe the pointer on the astrolabe. As a calculating machine it could perform measurements as well as dispense with the need for them, thereby helping to give even those people who were unable to count or measure a reliable indicator of the hours. Hermann invented just such a clock *c*.1050: the column-sundial. In one session on the astrolabe he compiled a table of changes in the sun's height during the year on the Isle of Reichenau, and converted their degree values into proportionate sections, using a geometrical (almost trigonometrical) figure; he transferred these to vertical half-monthly lines on a small cylinder, extended the lower ends to curves performing

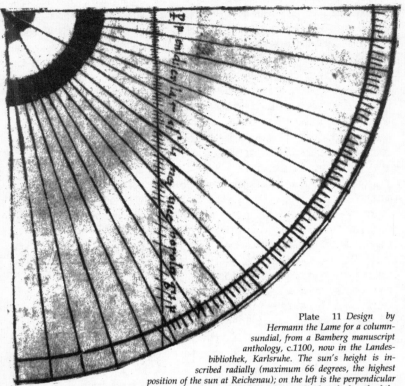

Plate   11 *Design   by
Hermann the Lame for a column-
sundial, from a Bamberg manuscript
anthology, c.1100, now in the Landesbibliothek, Karlsruhe. The sun's height is in-
scribed radially (maximum 66 degrees, the highest
position of the sun at Reichenau); on the left is the perpendicular
length of the column; between it and the middle perpendicular, the length of the
gnomon. On the middle perpendicular (perpendicularis linea) up to the degree points for the sun's
height are the shadow-lengths for the month-lines on the shaft.*

the same function as the hour-curves on the astrolabe, and at
the top of the cylinder installed a small rotating horizontal
gnomon. This was adjusted to the current half-monthly line,
and set according to the position of the sun; the hour of the day
was read from the point of convergence of the shadow and the
curve. The cylinder-sundial was cheaper to make, simpler to
operate and easier to move than the intricate astrolabe, and yet
within its narrow field of operation it was more accurate in
indicating the time than any conventional sundial or water-
clock. As a 'shepherd's clock', it served the particular uses of
all those who lived in one surveyable region, and whose work

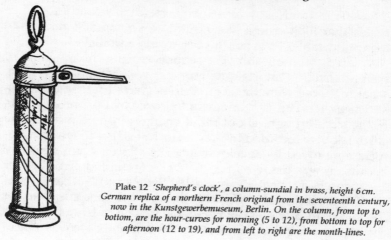

Plate 12 *'Shepherd's clock', a column-sundial in brass, height 6 cm.*
German replica of a northern French original from the seventeenth century,
now in the Kunstgewerbemuseum, Berlin. On the column, from top to
bottom, are the hour-curves for morning (5 to 12), from bottom to top for
afternoon (12 to 19), and from left to right are the month-lines.

time depended on the position of the sun.[99] This was the layman's method of determining times; computation, on the
other hand, was time-reckoning exclusively for scholars.

In the early 1040s Hermann used computation to improve
the martyrology of Notker the Stammerer. He recalculated the
feast day of the oldest saint in his abbey, St Mark's Day, and
used historical sources to correct dates in the life of the most
recent saint, Ulrich of Augsburg.[100] Liturgical custom was not
the same as historical truth – thus far Abbo was right. Neither,
however, was computistical calculation the same as natural
truth. In 1042, in his *Compotos*, Hermann asked, 'Whence
comes the error that the real age of the moon so often does not
correspond to our reckoning, *compotus*, or the rules of the ancients, and why, as Bede himself admits and our own eyes
confirm, does a full moon appear in the sky in most cases one
day, and in others two days, before the computed date?' The
sobering reply was that anyone wanting to get close to the
natural truth had to make more thorough observations and
reckon more accurately than Bede, and to pay regard to mini-
mal fractions on the abacus, and to tiny deviations of degree
on the astrolabe.

The *compotista* was first and foremost a *computator* and *calculator*. He did not look for basic formulae to help the mass of
lethargic and forgetful people, but instead sought the *naturalis*

calculatio examined by all experts. Hermann computed it in tables that looked less like figurative poems and more like logarithm tables.[101] When he began his chronicle in 1048 he calculated the thousand-year sequence using the new tables. He understood that in human time as well as in natural time, what mattered were those hitherto neglected fractions, atoms and *momenta*. The same moment followed by the precise orbits of the stars brought about rapid changes in human fate. The chronicle also recorded the key historical stages to which time-reckoning owed its development. The *compotus* had a history because it was made by people who committed errors. After Hermann had repeatedly corrected Bede's and Notker's asser-tions and continued to discover discrepancies, he suspected at the end of his life that the venerable system of Christian time-reckoning was based on mistaken assumptions. He was able to make this observation because he was the first person since Cassiodorus to place his trust more in the *experimenta* of time-measurement than the *rationes* of computation. If experts who relied on the astrolabe, followed his example, they would in future probably be able to agree on a more natural computus.[102]

This could not be expected to come about so soon. Hermann the Lame was the last person to master the entire spectrum of subjects between the computus, the martyrology and the chronicle. By the time he died in 1054 the bond linking these three genres since the Carolingian period had been severed. The behaviour of the priest-monk Adalbert of Benedictbeuern at this time is revealing. In 1047 he began compiling a book of the dead for his abbey and entitled it *istum computum*, because like ancient martyrologies, it was arranged according to days in the ecclesiastical year and included computistical tables. Unlike Bede, however, Adalbert no longer regarded the litur-gical ceremony with pious reverence. Between commemora-tive days for the dead he wrote the date of his own ordination – 13 March, 1047 – as if the martyrology were a novice-calen-dar.[103] As the history of the computus itself proved, the time of God and the saints did not tally with the time of individual human beings and peoples. Did, in fact, the Church and the world live in the same time?

8

Giving and Using Time in the Eleventh and Twelfth Centuries

In the late eleventh century the Gregorian reform of the Church and the Crusade movement precipitated the rationalization of European life by bringing divided nations together. However, the reformers regarded the present time within which they lived as the decisive period in ecclesiastical history, paying scarcely any further attention to history or indeed computation. Even the historical martyrologies were antiquated, since the religious revival altered the view of the saints. The popes used their religious authority to reserve the task of entering new names in the saints' calendar for themselves, and Usuard's handbook sufficed for more ancient times without antiquarian sophistry.

It was primarily the French, fellow countrymen of Gerbert of Aurillac, who believed that time and numbers found their way into everyday events more than into the history of salvation. They no longer strove to achieve Carolingian-style combinations, and consequently made their auxiliary disciplines independent of one another. Between 1081 and 1091, Master Gerland of Besançon was the last person to make use of both the abacus and the computus, yet he was not working towards any general, long-term objectives. He regarded the art of the *abaciste* as the most topical subject because it involved problems of daily economizing ('How do you share one hundred Marks among eleven tradesmen?'), and for counting purposes, he was already using the new Indian numerals and their

Arabic names. Gerland overlooked the fact that this skill could also be used for time-reckoning. He added a brief martyrology to his 'computus', but in the cold light of day even computists were no more than *calculatores*, studying, for example, the constellation of a solar eclipse that occurred in 1093.[104]

The reformist and highly educated clergyman Odo of Tournai wrote *Rules for the Abacus c.*1090, emphasizing the educational value of arithmetic for all sciences. The *Rules* sounded a different note from the writings of Cassiodorus and Agius. Odo held that without a number theory no *abacista* could understand the rules of reckoning, *calculationis id est computi argumenta.* The computus was annexed in full to the new arithmetic that reappeared with its Indian numerals and Arabic names.[105] Just as numbers became separated from time, so theology became free of history. Honorius Augustodunensis presented religious truths in a generally comprehensible way in his *Elucidarium* of *c.*1100. He did not set out to ascertain the year of the Creation, but rather the length of time God needed for it, concluding that this was 'a moment'. Satan, however, was 'not a full hour' in Heaven, and Adam was in Paradise for 'seven hours, no longer'. Honorius did not consider the year Christ was born, but rather the fact that the event happened at midnight like the Last Judgement, the date of which was of absolutely no interest to him. The fact that Christ lay dead for forty hours prompted a mystical multiplication rather than a computation of Easter. Jesus did this 'in order to bring to life the four quarters of the world that had died in the Ten Commandments of the Law'. Time was, in essence, neither the liturgical period nor the cosmic orbit, but the single moment. People in active work soon began to complain about lack of time. Number, however, was a function of allegory, an important symbol of that ineffable being that timelessly encompasses all of humanity.[106]

Augustinian historical theology, for which Hugh of Saint-Victor, a native of Germany, became the most influential spokesman, confined the comparison of times to these symbolic numbers. In 1130 Hugh drew broad analogies between the six world eras and the six ages of human life. The numerical framework he adopted for this purpose was Bede's; he did not deign to use it for computing shorter times. He had outgrown the mathematical interests of his schooldays, which might have been inspired by Odo of Tournai, preferring to use the astrolabe for measuring the earth rather than for determining times. When Hugh referred to the *compotus* he meant only a simple preparatory exercise for religious studies, namely enumeration of the 150 psalms that pupils were supposed to learn by heart so that they did not have to look them up every time in their books. Training the memory to retain numbers was useful for reckoning time as well as saving it, but only as a means to more subtle ends. Hugh regarded reading and writing as the epitome of higher education; indeed, he interpreted the entire physical world as a book written by God. Its symbols could be deciphered only by religious meditation.[107]

From the eleventh century the seven letters from a to g were used to represent musical pitches. In his treatise on music written *c.*1140, the reformer of Cistercian choral music, Guido of Eu, held that it was the sound b rather than the sign b that fitted into the *naturalis compotus litterarum*. Guido used *dispositio* to explain *compotus*, meaning the conventional sequence of letters in the alphabet, although they were not entered in tables for time-reckoning, as was the case with Abbo, nor did they distinguish the duration of the sounds.[108] The book-metaphor found its way into the nomenclature of the specialist literature. However, the audible communality between time-reckoning and church music that Hermann the Lame was still emphasizing made way for new specializations. They transposed the clerical rationalization, formerly begun by monastic reformers, to the daily lives of layfolk.

In Germanic countries the concept of time and the science of number became decoupled more gradually from their religious connection with worship and observation of the stars. Historians in the German empire continued to think of themselves as time-reckoners for half a century after Hermann. One was the Irish recluse Marianus Scottus who wrote a chronicle in Mainz extending as far as 1076. It included his autobiography, but also revolved around the disputed dating of the Passion of Christ; it denied Dionysius the distinguished title of *compotator* that it bestowed on Bede.[109] The secular priest Bernold of Constance resumed the chronicle of Hermann the

Lame, opening it in 1074 with computistical rules and singing the praises of his predecessor as a *compotista*. He was referring to this same resourcefulness when he praised Hermann as a *calculator* in 1093. Between 1074 and 1096, Bernold composed a calendar including martyrological, computistical and historio-graphical entries in equal measure.[110] In Bamberg *c.*1100, a whole group of monks and learned secular clergymen in Frutolf of Michelsberg's circle devoted themselves to detailed studies of music theory, chronicling and computation. They were particularly fond of describing themselves as *compotistae*. In Frutolf's own chronicle, however, computistical notes had no more than a secondary role in representing the fluctuating times of the past and the restless times of the present.[111]

Here, too, the long respite of the computists had come to an end. In a south German abbey, presumably before 1100, an anonymous scholar, writing about the universe and the soul, continued hiding behind Bede's name, but was almost like Plato in distinguishing between *compotiste*, who merely reckoned, and *philosophi* who explored nature and truth.[112] About 1100 the Benedictine author of the *Gesta Treverorum* was thinking along similar lines when he bestowed two honorary titles on a Jewish doctor called Joshuah: he was highly educated in natural science (*phisica ars*) and distinguished as a time-reckoner, *compotista*. Even in Trier it was no longer taken as read that the two disciplines were in accord with each other.[113]

In his chronicle from the 1080s, the Benedictine monk Sigebert of Gembloux protested that Dionysius Exiguus had failed to achieve the *ratio compoti*, and had wrongly computed the years following Christ's birth. Sigebert's *Liber decennalis* made a separate study of problems central to time-reckoning in 1092 but no longer used *Computus* as the title. He repre-hended the mistakes made by the ancients, but concluded resignedly that Dionysius's deep-seated errors survived even the most convincing counter-evidence brought by the *moderni*, including the most progressive *cronographus*, Marianus Scottus. All the arguments of the *compotus* were 'devised by human astuteness rather than demonstrated by natural truth'. While the computists were debating their different suppositions, the natural motion of the stars vanished into the mist, and with it God's immeasurable order of times.[114]

Meanwhile, shorter-term endeavours had also attracted the name *computus* in Germany. In his 'Ecclesiastical History of Hamburg', written during the 1070s, the former Bamberg pupil Master Adam of Bremen differentiated between the recording of past events and the prognosis of health. When doctors and mountebanks predicted a long life for one of their patients, Adam called this *calculare*. However, when tracing the dates of the archbishops of Bremen from the last two hundred years in the Corvey Annals, he described the work as a *compotus*.[115] By 'computus' he thus meant neither a chronicle nor a guide to the computation of Easter, but at least a history book structured chronologically by year-numbers.

The monastic reformer William of Hirsau, educated in Regensburg, had no interest in history. He knew how to determine long-range astronomical times by using the astrolabe and following the instructions of Hermann the Lame. In 1091 Bernold of Constance said that William had himself devised a *naturale horologium* modelled on the firmament and had solved many problems relating to the *compotus*. In his abbey in Hirsau, *c.*1077, William no longer kept the liturgical times of day and night, especially the moment of waking, by the first crowing of the cock. Instead, he ordered, as Abbo of Fleury had done, that the times should be accurately determined and indicated by water-clocks or, if these failed to work, by candles and observation of the stars. This method of measuring short spans of time could not be classified as computus, particularly as it did not lead to books on the subject. Nevertheless, when William stipulated that the corn-master of the monastery should keep an exact account of the annual harvest yield, this was in one instance called *computus*.[116] The word acquired overtones of measurability, standardization and the written form. In a Regensburg manuscript written in the early years of the twelfth century, and including works by Hermann the Lame, *conpotus* was simply Germanicized in a list of words to *zalpöh*. The word did not mean a book on time, merely a book full of numbers.[117]

The English were even more resolute in refashioning the computus. By 1086 the Norman kings had already formed an alliance with applied arithmetic, in the compilation of the

Domesday Book. Having resumed Bede's historical work in 1120, the Benedictine monk William of Malmesbury was familiar with the astronomical significance of the word *compotus* and resisted new theories that, like the astrolabe, could only give rise to astrological nonsense. But for the year 867 of the Incarnation of the Lord he told of a battle fought against the Vikings in which the Anglo-Saxon king fell, recording the toll of his countrymen's lives thus: 'nine dukes, one king, as well as men without number, *populus sine computo*'.[118] If losses of men in the war were now quantified more accurately than before, then this was even more true of monetary income in times of peace. The statute book of the Anglo-Saxon king Edward the Confessor, edited *c*.1130, made the City of London the permanent seat of the royal courts of appeal, including for *ardua compota*, controversial payments to the crown.[119] About 1131, King Henry I entrusted two profitable offices of administration to the citizens of London against the payment of 300 pounds *ad compotum*: in other words against arithmetical proof before the royal financial authority, the Exchequer.[120]

About 1178 the Treasurer, Richard of Ely, gave an account of this London authority's mode of operation and discussed not only the *calculator* who settled the sums in the old-fashioned way on the abacus in the Upper Exchequer, but also four *computatores* who counted the incoming cash in the Lower Exchequer. Richard described these annual final statements of account–the Pipe Rolls–as the *magni annales compotorum rotuli*. Financial accounting became the very historiography of the present.[121] The computus was on the verge of abandoning the theoretical and invariable explanation of temporal concepts and becoming closely submerged in current arithmetical practice. Would the new monetary system – the interest incurred on loans and the validity period of bills of exchange – transform the previously ecclesiastical and qualitative association between time and numbers into something material and quantitative?

9

Divided and Appointed Times in the Twelfth and Thirteenth Centuries

Twelfth-century Europeans did not succumb to the intoxication of numbers. The age of early scholasticism determined the subsequent fate of the computus by theoretically apportioning time. The new monetary economy caused offence, but this was not so much because it was based on quantity; kings and bishops, noblemen and peasants all behaved in a similar way. The provocation was due more to the fact that municipal money changers made God-given time, which belonged to no man, work for themselves. Teachers did virtually the same thing, selling the gift of their knowledge on the town market instead of imparting it within the confines of the Church in exchange for God's reward.[122] When rising groups of townspeople occupied long periods of time for momentary purposes, it was necessary to check what parts of this time rightfully belonged to them.

The philosopher Peter Abelard addressed this topic in Paris c.1140, in his central work *Dialectica*. His analysis of language was based not on the written letter, but the spoken word, and unlike Augustine he adhered to the grammatical division of language into the three tenses of past, present and future. Abelard declared this to be purely a human invention, and made a distinction between this linguistic time and natural time. According to Aristotle, it was only within this division that time and number fall under the category of quantity; it consisted purely of brief moments, *instantia* or *indivisibilia momenta.* Such moments cannot be combined linguistically until humans are overwhelmed by the desire to survey human triumphs and tribulations. Although we say an action or passion occurs 'hourly', 'daily', 'monthly' or 'yearly', forming an analogy with the course of the stars, basically human time does not arrange religious or natural relations, merely personal or social ones. It cannot be divided arithmetically, nor can it be calculated by computation, but it is as much ours as our physical and intellectual work.[123]

Abelard used the terms *computare* and *computatio* only when considering to which logical category he should assign a particular concept.[124] Here he was speaking metaphorically; by itself his linguistic logic did not touch upon natural science anywhere. 'For philosophy can do more than nature.' By his own admission, Abelard had a limited understanding of arithmetic, and he bracketed astronomy, as *mathematica*, with astrology, giving geometry merely a passing mention and music none at all.[125] Together with computistical studies, the ancient numerical sciences shifted to the periphery of scholastic hermeneutics and their nurturing ground, the university, with its faculties of theology, law, medicine and philosophy. Abelard himself confessed that its masters calculated no differently from usurers, collecting fees from their students for lessons measured by time. If numbers merely recorded banal payments, they warranted no further attention from the scholar, least of all in cases where he accounted for his time.[126]

The historians came around to Abelard's way of thinking. The two most prominent historians of the twelfth century, the German Otto von Freising and the Englishman John of Salisbury, saw themselves as literary chronographers – *cronographi* or *cronici scriptores* – and no longer as computistical time-reckoners. Although they venerated Bede, and attended to the sequence of years – *annorum supputatio* and *series temporum* – they circumvented controversial questions of time-reckoning as well as the word *computus*. Without fixing upon exact dates and numbers, they described a momentary fluctuation of triumphs and tribulations between the unmastered past and the uncertain future, rather than a succession of recurring and advancing years. Their contemporaries began to conceive historiography itself as a variable historical phenomenon rather than as an attestation of the eternal.[127] Henceforth, we no longer hear any of Europe's great historians declaring a passionate interest in computistical studies.[128]

The displacement of mathematics and natural science from the centre of Latin education had mixed consequences. What holy days and numbers lost in importance was gained by appointed dates and by calculations. Instead of proclaiming the reality of God, they formed relationships between people. The canonical law of the twelfth century, sworn to the leadership role of the papacy, revived the Carolingian stipulation that every priest should master the *computus*. Since clergymen were meant to have an all-round education, Gratian included it in his *Decretum*, c.1140. However, as he recognized arithmetic only as a field of limited insight like geometry and music, rather than as a subject of general piety, what was required was simply knowledge of the current rules.[129]

Computation thus came under the charge of jurists. To continue computing bordered on heresy, because people were now beginning to use pagan methods of reckoning. In the age of the Crusades Europeans reluctantly adopted them from their Islamic adversaries.[130] Anyone who refused to be deterred was freed from the burden of establishing global meaning now incumbent on the universities. He also overcame the cumbersomeness of Roman numerals, the abacus and counters. Furthermore, the Babylonian, Hellenistic and Arabic sexagesimal system of infinitely divisible minutes, seconds and terces, released him from the monopoly of natural, or possibly rational, numbers that had restricted the calendar since late Latin antiquity.

From the late twelfth century, time-reckoning underwent an unexpected revival that modern scholarship has scarcely taken account of, but which brought serious consequences in its wake. *Abacistae* had already toyed theoretically with Islamic methods of writing and reckoning in the late eleventh century. The new practice of reckoning by the decimal system, the algorithm named after al-Chwârizmis's book of arithmetic, was based on number theory and found its way into scholarly computation sooner than into royal fiscal policy and the bourgeois monetary economy. A start was made in 1143 in southern Germany by the essentially conventional 'Salzburg computus', preceded by an awkward introduction to algorithm. Without exploring the new possibilities arithmetically, Anonymus converted time-reckoning to Indian numerals, while historians and businessmen went on struggling with unwieldy Roman numerals for another two hundred years.[131]

The dean of Paderborn Cathedral, Reiner, fully exploited the opportunities provided by the decimal system in 1171 in his *Computus emendatus*, a moving and brilliant medieval scientific achievement that has since been forgotten. Its title was indicative of the fact that computists now no longer had to promulgate the eternal order of times, but constantly correct the order that had become erroneous. Using Indian numerals Reiner ascertained the important fractions more quickly than Hermann the Lame whose calculation of the lunar month he used as a basis. He proved that Dionysius Exiguus's *antiqua computatio* for the solar and lunar year contained errors amounting to one day every 315 years. His conclusion bordered on heresy: in order to fix the dates of Christ's lifetime it was necessary to employ the calendar used in Jesus' lifetime; that, however, was neither the later Roman calendar, nor indeed the present Christian calendar, but the ancient Jewish one. Time was now historically relativized by computists just as it had formerly been by historians.

The astronomical argument used by Reiner to extricate himself from this predicament sounded hardly more edifying. All methods of time-reckoning, the Jewish calendar included, produced no more than approximate values, while the 'manifold fluctuations in the moon's course' prevented even the *studiosus compotista* from establishing accurate dates. Sanctioned by Moses and consequently the oldest calendar in the world, the Jewish calendar had by this time been working faultlessly for 4,930 years; nevertheless, its principles did not correspond to conditions at the time of the Creation. This appalling discovery no longer affected just Dionysius, but also Bede and all universal chroniclers after him: was the world older than all the calendars? Reiner left his readers to answer this for themselves; his question was not taken up again until four hundred years later, by Scaliger.

In any case, the *compotus* no longer afforded insight into God's Creation of the world. It was now only 'the science we use to determine the days on which the feasts take place each year'. In order to eliminate at least the worst discrepancies in the Dionysian cycles, the Westphalian scholar was the first to undertake a pragmatic correction of the ecclesiastical calendar by recalculating the dates of the Crucifixion and Resurrection of Christ using the Jewish lunar calendar. As he knew, 'the Church does not condemn lightly something to which it has long adhered.' It was not the truth of Christian religion that depended on whether the Church modified its stance on the computus, but rather its damaged reputation *vis-à-vis* Mosaic and Mohammedan culture, its scientific character.[132]

Although cosmic, recurrent, natural time eluded human beings, they were at least able to co-ordinate their historical and unique social times. Against the background of the 'twelfth-century Renaissance' they preferred to follow ancient philosophers rather than biblical patriarchs. On a visit to Rome *c.*1200, an English scholar called Gregorius beheld the two Dioscuri of the Quirinal and, next to them, marble horses of wondrous size and beauty. As Roman tradition had it, the naked men were humble philosophers, and their horses symbols of political power harnessed by the spirit. The Englishman, however, was told they were memorials to earlier, probably ancient, *compotistae*, and were accompanied by horses because like these, time-reckoners were quick-spirited. Whatever the wily guide thought, the honest tourist knew so little about ancient methods of determining time that he no longer recognized the Vatican obelisk as a sundial, assuming the globe at the top to be Caesar's tomb. It merely reminded him of the presumption and transitoriness of secular power; this consolation was of more help to the ordinary citizen than learning how to reckon and measure time.[133]

Alexander of Villedieu, a Norman who collected a *Massa compoti c.*1200, doubted whether the intellectual effort was warranted. He placed time-reckoning in his encyclopaedia somewhere between the subjects of grammar and canonical law. As a mathematician he also mastered the new Arabic methods of reckoning, providing instructions on their use in his own *Algorism*. Unlike Helpericus of Auxerre, however, he no longer drew comparisons between *ars calculatoria* and *compotus*. In approximately five hundred verses he laid down what clergymen were supposed to know about time-reckoning, and what they no longer really knew. They could learn the rules of his perpetual calendar by heart and apply them mechanically, without having to do much reckoning or ask many questions. Of particular help to them were the Golden Numbers from 1 to 19, representing the nineteen-year cycle and fixing the Easter Day of each calendar year. Caesar had purportedly invented them himself. It was as if he had wanted to celebrate Christ's Resurrection in an appropriate way . . .

Alexander introduced a scholastic classification of time-reckoning corresponding to the late Carolingian distinction between the 'large' and the 'small' computus; it circumvented Reiner's demand for reform, and was repeated for centuries.

> *Compotus* is the science of differentiating times by foolproof and rational means. It is called *compotus* from *computare*, not because it teaches us how to reckon, but because we learn it by reckoning. It should be noted that it consists of two parts, the philosophical *compotus*, and the vulgar or ecclesiastical one. The philosophical *compotus* is the infallible science of dividing up time, *scientia temporis discretiva infallibilis*, and does not concern us. The vulgar, or ecclesiastical, *compotus* is the science of dividing time according to the custom of the Church, *scientia temporis discretiva secundum usum ecclesiae*, and it is this *compotus* that we wish to address.[134]

Alexander might well have had the title of Abbo's *Computus vulgaris* in mind, but for him these words meant the opposite. If time no longer expressed God's eternal truth, clerics did well to avoid risky calculations and confine themselves to what was conventional; for the purposes of everyday living it was probably the most useful thing to do.

In the thirteenth century the question of whether stipulated social time ought to, and could, be synchronized with computed, physical time, remained a subject of debate. In his *Computus ecclesiasticus* of 1232–5, the Parisian university teacher John of Sacrobosco reached the same conclusion as Alexander of Villedieu, whose verses he laboriously quoted, and for whose views the very title of his book declared support. John also wrote widely used textbooks on astronomy and arithmetic, and was initially more active than Alexander in shielding the subject of time-reckoning from general education, designating it as a specialist area of astronomy: 'Computus is the science of observing times by the movements of the sun and the moon, and comparing these with each other.' At high scholastic universities, astronomy dealt almost exclusively with speculation, scarcely with observation, and made use of the astrolabe only as a teaching aid in cosmology, not for correcting the calendar. The sort of astronomy consisting merely of reckoning was no use for this purpose. Although it calculated the movements of all the heavenly bodies down to the last detail, computation formulated only very rough temporal equations by the solar and lunar cycles.

Employing the methods of Ptolemy and the Arabs, John was able to calculate the movements arithmetically down to small astronomical time-units such as minutes and seconds. Like Reiner of Paderborn, he noticed that the results deviated from the assumptions of the ecclesiastical calendar. But it was only in the case of the tropical solar year that he proposed a modest correction: the 'calendrical order' could be reproduced by omitting the leap-day every 288 years. As far as the synodic lunar month and thus the computation of Easter were concerned, John rounded up the figures 'according to universal custom', and took refuge behind the purported resolution of the Nicene Council in 325: 'Since the General Council forbade any alterations to the calendar, modern scholars have had to tolerate such errors ever since.'[135] It was possible to live with these errors, and John's work became the most popular university textbook on time-reckoning. Luther's ally Melanchthon republished it in 1538 for the recently founded University of Wittenberg.

In the 1250s the greatest encyclopaedist of the High Middle Ages, the Dominican Vincent of Beauvais, expressed full appreciation of the progress made since Isidore in dealing with numbers. He called for Indian numerals to be used in time-reckoning, discussing them both in the same section entitled *De computo et algorismo*. Also, in keeping with the times, he was familiar with two meanings for the computus: for him *computare* in its broader sense meant the same as *numerare*, hence the *computus* of each process of enumeration. In its narrower sense, *computus* merely differentiated times by the courses of the sun and the moon in order to determine the movable feasts of the calendar. Alexander of Villedieu had written in a similar vein. Vincent added that the new numerals were needed to work out the requisite small units of time and long numerical sequences. He then waved this aside, maintaining that this was a task for specialists and not suitable for discussion in an encyclopaedia. In the historical part of his work Vincent himself dispensed with computistical methods, preferring to fix dates relative to actual names of rulers, like a chronicler (ultimately in the spirit of Herodotus), rather than by an absolute number that he found too uncertain.[136]

Was time-reckoning, then, no longer related to those Christian doctrines comprehensively analysed by high scholastics? Its leading lights, the Dominicans Albert the Great and Thomas Aquinas, well knew that time could be computed and observed with reasonable accuracy using modern methods, and it was almost with embarrassment that they called the conventional and approximate procedure *computus ecclesiasticus.* It was not their task to make it more accurate by using the Islamic astrolabe, or indeed of reforming it into the *computus philosophicus*; besides, this would also have conflicted with their philosophy. Following Augustine's example, they subjectivized time, refusing to concede that it was completely real in Aristotelian terms.[137] Therefore, not unlike Abelard, they presented theoretical reasons for excluding time-fixing from the theoretical sciences, transferring it to the realm of technical practice.

10

The Confusion and Management of Calendars in the Late Middle Ages

Between 1263 and 1265 the Oxford Franciscan Roger Bacon railed against this lax opinion of the book scholars in his epoch-making *Compotus*. His definition of the subject, following that offered by Robert Grosseteste, sounded like a declaration of war against John of Sacrobosco, and one of support for Reiner of Paderborn. At the same time, it brought together scholasticism's increasingly fragmented notions of time in a new way.

> The science of time is the science of differentiating and enumerating times that arise from movements of external bodies and human laws. The authors call this *compotus*, after *computare*, because it teaches us how to compute time by means of its parts. This division and designation occurs in three ways; in the authors' *compoti* some things are designated by nature, some by authority, and some merely by custom and caprice.

On the basis of nature, distinctions were made between years, seasons, months and days; according to authority, the natural year was distinguished from the civic year, the solar month from the lunar month; and on the basis of custom, distinctions were made between months consisting of 28, 30 or 31 days.[138]

Bacon's threefold division stemmed from Bede, but he altered the emphasis. Time could be more than a mixture of contradictory signs that we take from books; its resubstantiation would change the reality of our lives. If we Christians did not wish to look ignorant before Muslims, we must be sensible in dividing up and exploiting the short time we had before the descent of the Antichrist. Anyone recognizing mathematics as being fundamental to the shaping of our world would also apply science to daily life in order to rationalize civic trade and perfect the Christian way of life, and would not arbitrarily round numbers up or down.

Our Easter feast is now three or four days out of step with the moon. We can outwardly reconcile nature and art, even if – indeed precisely because – we can no longer work with whole numbers in the same way that earlier *compotiste* did. The solar year and the lunar month cannot be slotted into simple arithmetical equations, and hence the course of the stars and the ecclesiastical calendar can only be roughly approximated. Time-reckoning was still so far and away superior to time-measurement that Bacon neither undertook nor proposed any experiments involving instruments, neither the astrolabe, which was far too inaccurate, nor the Islamic water-clocks and sundials of the High Middle Ages, which were much too intricate. But the Christian division of time was at least supposed to correspond as closely to celestial phenomena as it had in the seventh century, when the Arabs discovered the best possible calendar. Christian *compotiste* had to correct six hundred years of mistakes that have left the computed dates for the spring equinoxes and the first full moon increasingly out of step with observed points in time.[139]

Everyone could see that Bacon's criticism was apposite. But who was supposed to bring about the change he was demanding? Ordinary mortals lacked the education and time needed to grasp the time-reckoners' argument, let alone make a decision about it. And why should they, as long as popes and kings prescribed to them the cycle of the years, the dates of feasts, and the structuring of days? Calendrical problems were more often solved by judicial decree than by computistical research. Knowing this, Bacon made his appeal for a reform of the calendar to Pope Clement IV in 1266.[140] But did the papacy have enough scholarly conviction, political courage and social authority to reduce the whole of Christendom to a common denominator?

The papal Curia initially saw no cause for spectacular measures, for the very reason that it wanted to standardize time legally. In 1286 the Frenchman Guillaume Durand, having entered papal service under Clement IV, and having since risen to bishop, devoted the whole of *De computo et calendario*, the last volume of his handbook on liturgical law, to the prevailing norms of time-reckoning. However, he defined *computus*, not as a matter of merely stipulating or regulating time, but as 'the science of ascertaining time by the course of the sun and the moon'. After this concession to the objectivity of time-division, Durand fell back on Alexander of Villedieu, contrasting the vulgar, or ecclesiastical, *computus* with the astronomical or philosophical one on which he wished to spend no further time. Although he made repeated reference to the *error nostri computi*, and sought to correct it in detail by adding further rules, he was not contemplating a general reform of the calendar; and on Bacon he maintained an eloquent silence.[141] This was the last attempt by an expert to preserve the traditional method bearing the title customarily used since Cassiodorus, *De computo*.[142] Instead, experts devoted more and more of their attention to the theme addressed by Bacon, *De calendario*, and challenged traditional time-reckoning.

This same tendency characterized changes in neighbouring areas that largely took away decisions about time and numbers from the experts. Astronomy and cosmology became less necessary for determining times primarily because in the late thirteenth century the astrolabe lost its pre-eminence as the most accurate device for measuring short time-spans to an old rival, the solar quadrant. The solar quadrant had been introduced from Islamic Spain together with the astrolabe in the eleventh century, but was initially less useful than the astrolabe. The quadrant, comprising a quarter of an astrolabe's reverse side, reliably converted the height of the sun in equatorial zones into hours of the day, and when fitted with a movable calendrical scale it was meant to do the same for medium latitudes. There, however, the device was out of step with true local time by a full hour, especially at noon on the longest day, as it read the time from parallel lines rather than curves such as those appearing on the front of the astrolabe.

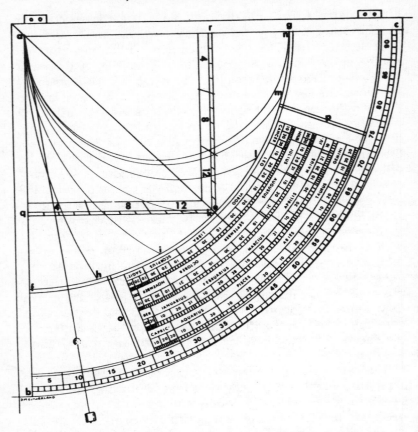

Plate 13 *Solar quadrant, redrawn by Nan L. Hahn from the manuscripts of John's Manual, Montpellier, c.1280. At the top is the sighting device, at the bottom the pendulum; on the quadrant is a shadow-square (q–r), six hour-curves (h–n), and a movable calendrical scale (o–p), here set at the highest position of the sun at just under 70 degrees, corresponding to the latitude of Montpellier.*

Hermann the Lame had with good reason adopted its hour curves for his column-sundial. Sacrobosco, however, was familiar only with the inaccurate line-quadrant that scholars call *quadrans vetustissimus*.

The line-quadrant was subsequently adapted to complex hour-curves, apparently by Englishmen living in southern France, a certain Guillelmus Anglicus living in Montpellier or Marseilles in 1231 being the first to know about it. A treatise on the new instrument was published in Montpellier in 1263, with a revised version following shortly before 1284, again in Montpellier, written by a certain Master John who may also have come from England. His work spread swiftly throughout Europe. As the new quadrant (called *quadrans vetus* by scholars) occupied only a quarter of a circle it could be made larger than the astrolabe or the column-sundial, thus giving the time of day more accurately. Its larger size also made it more suited to practical purposes than the astrolabe, especially to geometrical and geodesic measurement. Ultimately, all the astrolabe's astronomical and cosmological features were omitted, allowing even non-specialists to measure the time of day accurately to the nearest quarter of an hour.[143]

In the meantime, in a commentary written in 1271 on Sacrobosco's major astronomical work, another of Bacon's fellow countrymen, Robert Anglicus, ventured a bolder initiative. When considering how to enumerate the hours of day and night he failed to accept that all instruments for observing the stars and noting the hours were tinkering in a roundabout way with the 'artificial day' and its unequal temporal hours, whereas the 'natural day' with its twenty-four equal equinoctial hours could have been handled far more quickly and accurately: 1 hour = 15 degrees of the circle. Anyone constructing astronomical devices other than on the basis of astronomical findings should not have been surprised at the perpetual errors that followed. 'It is impossible for any *horologium* to attain astronomical accuracy. Constructors of *horologia* (*artifices horologiorum*) are trying to make a circle (*circulus*) which will move just like the equinoctial circle; but they cannot quite complete their task. If they could, it would be a really accurate *horologium*, and would be more valuable than an astrolabe or any other astronomical instrument for noting hours.'

The problem could have been solved mechanically if Vitruvius's ancient water-clock had been used as a model. It had been further developed in Islam, and in the Latin West it was similarly called *horologium*. Was it not possible to release it from its dependence on cumbersome water-power and thus adapt it more readily? Robert thought in terms of a tared and freely oscillating wheel moving regularly by means of a lead weight at its axis, completing one revolution from one sunrise

Plate 14 *Medieval water-clock, from an illustrated Bible, Paris, c.1250, now in the Bodleian Library, Oxford. The Prophet Isaiah shows the fatally ill King Hezekiah a* horologium *(2 Kings 20:11). Small bells are positioned above a wheel consisting of fifteen metal cones with centre holes; below, a water basin but no dial or illustration of the water supply.*

to the next, plus the one degree by which the tropical motion of the sun lagged behind the sidereal rotation of the fixed star. The dial could retain its tried and tested form, resembling the front of the astrolabe, whilst the 360-degree scale on the outer rim could be divided easily into twenty-four hours; the back, with its sighting device and calendrical scale, would become obsolete.

What a relief that would be to everyone! The mechanism would run virtually maintenance-free in all weathers, and would accurately reproduce the current position of the heavenly bodies. Experts would no longer have to carry out laborious measurements on the back of the astrolabe each time they required a time-check, and laymen would be able to gauge the hour correctly without miscalculating, and without damaging a precious instrument. Robert immediately linked the uniformity of the machine to that of the hours indicated and in doing so, described the very principle of the mechanical clock itself more accurately than was later done in practice. When this would happen depended on just one, albeit difficult, technical facility at which the makers of *horologia* continued to labour and of which Vitruvius had made no mention: the escapement releasing the wheel and the weight at intervals and preventing a premature wind-down.[144]

The attention devoted to simplified and yet more accurate time-fixing was also devoted to the applied and similarly sophisticated science of number. Even this science coped increasingly well without educated mathematicians and computists. By the time the Church and universities of the High Middle Ages had dismissed their use as something profane, numbers had become so bound up with the secular present at so many levels that clergymen were no longer able to untangle the threads. Experts in arithmetic were no longer the only ones demonstrating a numerical consciousness. From the thirteenth century, probably in Italy initially, anyone observing and recording the history of the period in which he lived would assiduously practise the realistic enumeration of people and objects and the plausible dating of events and changes.

In the same period a commercial rationality began to spread outwards from Italy. About 1250, *conto* in Italian still meant astronomical time-reckoning, like the Latin *computus* from which it derived. In the early 1260s the Florentine Brunetto Latini wrote an encyclopaedia, not in his local dialect, but in French. In this encyclopaedia computistical time-reckoners were still termed *conteour*, and their results *li contes de la lune et ses raisons*.[145] *Computus* and *ratio* were formally translated, but their meaning remained constant. Both amounted to abstract 'reckonability'. In the 1280s, however, Dante Alighieri wrote a collection of love poems entitled *Il fiore* in which *conto* was used to mean a relationship between two lovers. The metaphor did not revolve around the relationship of constant love to the passage of time but was aimed directly at reckoning and balancing income and expenditure, at economic accounting.[146]

Together with these procedures, the word *conto* found its way into the languages of neighbouring countries: into French as *compte*, Spanish as *cuento*, and German as *Konto*.[147] The pa-

pal chancellery helped complete the change to Latin, and by
the 1250s at the latest it had created the office of the *taxator* or
computator. The *taxator* had to fix and register the charges for
papal bulls, but did not have to compute feast days as Durand
had done.[148] In France the monetary economy also found its
way into public administration just as it had at an earlier point
in England. From the mid-thirteenth century the royal *curia in
compotis* assembled in Paris; it was soon devolved as a *camera
compotorum*. From 1304 it was called the *Chambre des comptes*, or
counting chamber.[149] In the fifteenth century, the Germans
borrowed the word *comptoir* to mean 'money table', 'writing
room' and 'trading place'; it still survives as *Kontor*.[150]

The financial world had to fix calendar days clearly and add
them up quickly. It was consequently no longer able to describe the date in our example as the sixth day before the
Nones of March, like the Romans. Anyone continuing to name
it after Christian saints would become caught up in the confu-
sion of regional cults. In Flanders, 2 March was dedicated to
the late Count Charles the Good, and in Bohemia to the late
Princess Agnes. Even before this, the chancellery of Emperor
Henry VI had begun to enumerate the days of the month as the
Normans had already been doing for some time. In 1191 a
Milanese decree from the emperor was issued on *quarto mensis
Decembri* (*sic*), 'on 4 December'. In principle, however, not even
Italy abandoned the ancient custom. In German-speaking re-
gions the constant enumeration of days in the form of '2
March' first appeared in 1252, in the municipality of Lucerne,
and for a long time failed to gain acceptance.[151]

On the contrary, from the late twelfth century, a new
mnemonic for dating by saints' feast days began to spread
outwards from Germany. This was the so-called *Cisiojanus*,
an abstruse-sounding poem consisting of twelve double
hexameters, adapted from computistical mnemonic rhymes
composed in the early Middle Ages. Each day of the month
was represented by a syllable. *Ci* stood for *Circumcisio Domini*,
1 January; the following two syllables *sio* represented 2
January and 3 January without feasts; *Janus* was reminiscent of
Januarius and of the fourth and fifth days of the month. On the
other hand, no saint's day was found for 2 March – just the full
syllable *ti* in *Martius*. This was nevertheless a practical method

of enumerating calendrical time without the mortifying precision of reckoners and clerks, and was thus subsequently translated into the vernacular.[152]

The fact that in Germany the year more often than not began with Christmas, in France at Easter, and in Italy and England with the Annunciation, was also a hindrance to business. Like the martyrological and computistical year, the civic and economic year now began on 1 January; yet even those who could tell this from the *Cisiojanus* did not convert their calendar. As with coin-minting and units of capacity, so it was with dating: in the midst of progressive fiscalization, attempts at rationalization merely increased the confusion. Many people had by now learned to count and reckon, but not all of them were enamoured of exact dates. People could live more leisurely in a world of approximate values, and if appointments were made, these were human appointments, in other words, flexible ones. The ability to read and write was unnecessary for remembering key points in time that had long since passed, even if inquisitive authorities were beginning to inquire about the dates of people's lives with increasing frequency. But many laymen were reluctant to have the details of their lives recorded in files and their enumerated time.[153]

Even those who viewed the world like a book, however, did not leave its fate to small numbers. In Germany *c*.1205, probably in Bamberg, a clergyman believing in the precise wording of the Bible hit upon the semi-cabbalistic, semi-algebraic idea that the length of the *saeculum* on earth was concealed in the Latin alphabet. Each of the twenty-three letters corresponded to a full century; the sequence began with the founding of Rome rather than the Creation, and rightly so, in that numbering by *saecula* actually originated in ancient Rome. Three letters, x, y and z, denoted that three hundred years were still to pass. The agitation surrounding the Hohenstaufen emperor Frederick II left many contemporaries fearing a much earlier end to the world. However, in 1288, after the commotion had subsided, the Cologne Canon Alexander von Roes returned to the reckoning used by the Bamberg cleric, with the result that the last *annorum centenarius* would be delayed until *c*.1500. He ultimately experienced doubts: other people anticipated a future lasting several thousand years, and followers of

Aristotle used 'natural arguments' to support their view that the world was everlasting. The safest course was to be prepared for everything at all times, and not wholly rely on any one moment.[154]

Why, in the face of such uncertain prospects, was there such a hasty attempt to standardize the present? Why regulate the year and the day with more precision than the pound and the penny, the mile and the foot? It was difficult to implement even the simplest standardization. The secular priest Elias Salomon from Périgord complained about this in 1274 when dedicating a *Scientia artis musicae* on the practice of music to Pope Gregory X. He argued that what he had found to his liking in the papal chapel of Avignon, but had not at that time been introduced everywhere, ought to be generally prescribed: the pages of choirbooks ought to be numbered regularly so that, when given references, singers could quickly find the correct page. Elias called this page-numbering *computus*, without establishing a connection with sequences of notes and rhythms or with divine worship and the ecclesiastical year. Higher education still depended on books. But Elias complained that many choirmasters were erasing these new numbers because they wanted to avoid making the task too simple for their choirboys. Book scholars often baulked at the intrusion of Indian numerals into writing; that aside, they did not always apply their knowledge to making life easier for others.[155]

When reckoners could nevertheless bring themselves to do so, how did they best serve their fellow men? Not by computing the distant end of the world, but by forecasting the weather for the following day and the harvest for the following year. An astrological manual, perhaps written by John of London, but compiled there in 1296, and revised *c.*1318 by Richard of Wallingford in Oxford, noted simply that at the time of the *fundatores kalendarii Romanorum* the winter solstice had fallen at Christmas, but that it now occurred more than eleven days earlier, as had been established *per sapientes compotestas in isto novissimo tempore speculantes*. Whether the author had Bacon in mind or not, he scarcely made any distinction between speculating computists on the one hand, and astrologers and meteorologists on the other. Indeed, he was not even concerned about reforming the calendar; in order to correct it he added three tables that were meant to apply to 'all times, past and future'. However, these tables went back only as far as 1176, and progressed only up to 1416, and applied solely to London. The fact that fourteenth-century Europeans loudly proclaimed quantification as their ideal and yet failed to practise it had more to do with their practical sense of relativity than with any theologically motivated despondency.[156]

The more minutely daily life was regulated, the more seriously the popes took their responsibility for the long-term future of Christendom. In February 1300, urged more by the Church than by personal impulse, Boniface VIII was robust in providing for this future, promising a complete indulgence to all Roman pilgrims for the current year, and immediately declared that it could be acquired 'in every hundredth year thereafter'. The papal proclamation of the 'Holy Year' was primarily intended as a celebration of the anniversary of Christ's birth, but it also revived the memory of secular feasts in ancient Rome, and promised the world a long future under the gracious rule of the papacy. In order that the faithful would not have to wait an entire lifetime or even longer for this blessing, in 1343 Pope Clement VI shortened the period that had to lapse before the next jubilee year to fifty years (according to Leviticus 25:11), thus allowing the Church to celebrate an *annus iubilaeus* again in 1350. This marked the beginning of the rapid growth in jubilee years culminating in our century, when we have merely come to celebrate our collective loss of memory.

The same Clement VI, a Benedictine scholar, also made provision for the more immediate future. In Avignon in 1345 he pressed for an improvement in the calendar. The same year, the Church celebrated Easter a whole week too late in astronomical terms. The Pope solicited proposals for improving the calendar, the most important of them from the Parisian university scholar Jean de Meurs who had long emphasized the connection between time and number, and more vigorously than high scholasticism. In his *Notitia artis musicae* of 1321 he had unequivocally declared his support for Aristotle's theories: *Est autem tempus mensura motus*. Two different forms came together in this eventful, calculated time: in the musical recital, for instance, natural caesurae marked a beginning and an end, and mathematical caesurae subdivided the whole. As Jean explained in his *Arithmetica speculativa* of 1324, numbers fell into two categories: natural ones visible in objects, and mathematical ones abstracted from objects.[157]

In his *Epistola super reformatione antiqui kalendarii*, Jean explained to the pope that there was only one mathematical solution to the problem of the calendar. This solution was facilitated by the fact that c.1272 the court astronomers of the Castilian King Alfonso the Wise, prompted by Islamic examples, had recorded the movements of the sun and the moon in tabular form with more precision than before. Jean, who had been correcting and commenting on the Alfonsine tables since 1318, now had to reconcile two incommensurable numerical sequences, each arithmetically complex in itself, and each constructed by an astronomically different method. The tropical solar year could be computed by omitting leap years in such a way that the spring equinox would again fall on 21 March, as it had done at the time of the Nicene Council. By adding to the Golden Number, the synodic lunar month could be moved so that Easter Sunday would again fall immediately after the actual spring full moon. For both, however – and this was worst of all – it was necessary to shorten the calendar year by several days. Jean feared this *reformatio* might lead to quarrels about payments and contracts in the princely courts, and might even give rise to disturbances among the people. Accordingly, he merely altered the arithmetical formulae used by the experts, thereby merely providing a cure for inconspicuous symptoms.[158]

Others emulated him. Shortly after 1360 an anonymous scholar from southern Germany drew up a *compotistica figura*, a table proving – allegedly *infallibiliter* – not only the dates of the fixed great cycles, but also the movable dates of the *computus ecclesiasticus*. The author, however, immediately added rules permitting corrections to be made in an oddly schematic way to the table for leap years, Sunday letters, and Golden Numbers. 'If the basic number agrees with that given in the figure, all well and good, but if it does not, then correct it.' The reader was not told what the purpose was of carrying out these cross-checks.[159] Late-scholastic natural philosophers

like the Oxford *calculatores* were by now able theoretically to
quantify and modify short time-spans and infinitely high
speeds using algebra. But no one believed that measurements
other than practical empiricial values would produce accurate
results in the long term. *Calculare* did not, after all, mean to
check measurements and experiment with existing things, but
rather to calculate, and speculate about, uncertainties.[160] The
reason why the reform of the computus around the mid-four-
teenth century failed was thus not the despondency of science
nor the obstinacy of the papacy; its failure was due to the
prevailing notions of time and number in Christendom.

11

Mechanical Clocks and Rhythmic Differences in the Fourteenth Century

The prevailing notions of time and numbers were little altered even by the invention of the mechanical clock, whose revolutionary influence tends to be overrated by modern scholars. The clock was the earliest machine for measuring time; by linking enumeration with measurement, it overturned the old order of rank in the way Hermann the Lame had already urged. It combined the principles of the abacus and the astrolabe, a 'digital' running gear that counted backwards, and an 'analogue' dial that measured continuously. However, the fact that the clock did not immediately revolutionize people's consciousness of time and numbers is evident from our ability to date the invention only vaguely between 1300 and 1350, and the fact that no contemporary scholar is able to name the inventor.[161]

The inventor was unable to cause much of a stir so long as the invention of the escapement remained merely a realization of Robert Anglicus's proposal of 1271, and when even this went only half-way. The new machine was by no means meant to overthrow the old temporal order. It was progress enough if the astrolabe, mechanized in the style of a water-clock, needed readjusting only once in the morning and once in the evening; it would then be as accurate as before, indicating the uneven temporal hours of the following day and the next night that still dominated life and could be read from the curves on the inserted plate of the astrolabe. Specialists no longer had to

Plate 15 *Tower-clock from St Sebald, Nuremberg, late fourteenth or early fifteenth century, now in the Germanisches National Museum, Nuremberg. Height 43 cm, sixteen-hour dial, hour-studs for touching in the dark, and a small alarm for the warden who would then strike the bell by hand.*

carry out laborious measurements by day and night for each time-check, and laymen no longer needed to use their hands to find out the hour, but simply their eyes, and at night even just their ears. The fact that the new machine was given a striking mechanism, thus assuming the additional function of a bell, did not fundamentally alter people's sense of time. Although shorter measures of time than the seven canonical hours and the twelve temporal hours were now signalled from the tower, it was initially still the bellringer who sounded them by hand as soon as the striking mechanism woke him.

Nevertheless, Robertus Anglicus's central idea caught on because the machine almost automatically awarded priority to the uniform equinoctial hours. Anyone averse to adjusting the clock twice a day only had to wind up the mechanism periodically, as soon as he converted the hands and striking mechanism to the 'even' twenty-four hours corresponding to the full circle on the outer rim of the astrolabe. They were favoured by specialists as 'natural' hours, and laymen found it easy to read the time from the machine's dial. However, all its technical and scholarly benefits would have had little effect if the change had not also struck a chord with the mentality of town-dwellers. Their daily work, which was increasingly timed by instruments and rewarded with payments of money, was meant to be calculable and controllable within the town walls and hence uniform; there consequently had to be a common clock for employers and employees alike.

For these reasons, clock-hours of equal length gradually came to replace horary prayers as the unit of time, and Bede's recommendation became firmly established. In the fourteenth century the German word *Uhr* was still borrowed from the Latin *hora*, and more directly from the Italian *ora*. Prior to 1383 the people of Nuremberg had mounted an hour bell in the tower of the Sebald Church that the bellringer operated by hand. When it had to be replaced in 1396, its successor was called *Orglogck*, and was combined with a mechanical clock. Like the people of fourteenth-century Nuremberg, the Germans of today still refer to this timekeeping device whenever they say it is '18 Uhr' ('6 p.m.').[162]

The fact that the instrument could be seen hanging in the church tower and could be heard striking the hour meant that time was standardized only within the horizon of the church tower. Hours of the day were counted differently from place to place, in small hours, large hours, or whole hours, and it was rare to start the numbering at midnight as we do today. Nevertheless the mechanical clock severely disrupted people's consciousness of time in the late Middle Ages by exposing the synchrony of non-simultaneous things. It did not create 'modern times', and it certainly did not create a 'universal time', as is enthusiastically claimed by believers in progress. It blocked Bacon's synthesis and encouraged at least four conceptions of time with considerable rhythmic differences between clocks. According to all four, the clock was a symbol for a measured way of life in the midst of chaotic circumstances. However, while the symbols on the dial could be read more quickly than the letters in the book with which scholars had previously compared the world, it was no easier to interpret them.[163]

The two non-scholarly forms of symbolization preached humility. The German Dominican Heinrich Seuse, with his *Horologium sapientiae* of 1334, can be seen as representative of the first, spiritualized time of mysticism. The divine mercy of the Saviour appeared to him in a vision in the form of an elaborate clock whose melodious bells chimed every twenty-four hours. The mechanical clock and the carillon came to be seen as mirroring the soul. By observing the Passion of Christ at all times, a whole life long, the soul could be reawakened to eternal wisdom and raised above all external time in a moment, 'in a trice'. The God-loving soul apprehended this inner time in a similar way to Augustine, but shared it only with the Lord, not with the community of the Church or that of the town.[164]

A second, personalized concept of time had its roots in skilled trade, and centred on the hour-glass which, like the mechanical clock, first appeared in the fourteenth century. In the earliest illustration, painted by Ambrogio Lorenzetti in 1338 in Siena town hall, *Temperantia* is holding aloft an hour-glass. From the time of Isidore of Seville onwards, *Tempus* had been associated with this virtue, and the hour-glass was therefore especially suitable as a symbol of moderation, regularity, and satisfaction contained within the moment. It made working people aware of passing moments, silently and without a

Plate 16 *The oldest picture of an hour-glass, held by Temperantia. Painting by Ambrogio Lorenzetti (detail) in the Hall of Peace in the Palazzo Pubblico, Siena, 1338.*

numerical rhythm. Each individual divided and occupied such moments differently, the scholar in the study, the preacher in the pulpit, the advocate in the courtroom, the sailor on watch, the housewife at the oven. Yet in the hands of Death the hour-glass reminded them all of their final hour, and urged them to make use of the moment for as long as there was still time: 'Your last hour is one of these.'[165]

Two scholarly theories preached pride. A third, atomized concept of time adopted fractions of hours that previously could have been calculated, but not represented. The tower clock now struck the half- and quarter-hours too, and people thought in terms of minutes and seconds that up to then had been used only by astronomers. Could the influence of the planets on human destiny be reliably ascertained at last, as Firmicus Maternus had demanded? About 1330, the Oxford mathematician Richard of Wallingford, promoted by then to Abbot of St Albans, not only constructed a planetary mechanical clock, but also cast the horoscopes of the small children of the royal family, predicting their entire future beginning at the cradle. He found many imitators.[166]

The natural scientist Nicole Oresme can be seen as representative of the fourth, mechanized concept of time, that of late scholasticism. In 1377, in his *Book of the Heavens and the World*, written in French, Oresme described the universe as an *horloge*, a regular clockwork that was neither fast nor slow, never stopped, and worked in summer and winter, by night as well as by day. He drew a direct comparison between the movements of the planets and a mechanical clock that balanced out all forces by means of its escapement. 'This is similar to when a person has made an *horloge* and sets it in motion, and it then moves by itself.' In particular, the planetary clock came to represent the cosmos, the improved astrolabe rather than the accurate time-measuring device; its designers could compare themselves to the creator of the universal machine.[167]

Oresme's objection to the astrologers was the same as the lesson learned by the computists, namely that planetary movements were incommensurable with one another and thus never reconverged to form identical constellations. The clock's dial and hand movement, however, visibly confirmed the Aristotelian definition of time as the numerical value of a motion from a prior point to a later one. When Oresme claimed he saw a clock in the heavens, he had in mind the large mechanical clock that King Charles V had installed in his palace in 1362. From 1370 onwards, all Parisian church clocks had to keep in step with its somewhat capricious chime; it allotted townspeople their working day. It was the king, the designer *par excellence*, who stipulated how the course of social time was to run.[168]

The modern time-system consisting of man-made symbols

was already fully developed by the end of the fourteenth century. However, Europeans were less inclined than ever to divide their days on earth by a common denominator. The reforming councils of the early fifteenth century made a new attempt at this after papal endeavours had failed to overcome the schism. While one council, the Nicene Council of 325, had established the previous temporal order, it was a new council that had to put it straight and secure the future for the re-unification of the Church. These assemblies of all the clerical, political and scholarly heads of Christendom thus wanted an improved system of time-reckoning based on accurate time-measurement. In 1417 Cardinal Pierre d'Ailly presented an *Exhortatio super correctione calendarii*, written in 1411, to the Council of Constance with a satirical play on words, suggesting that great men of earlier times had devoted more care to the *calculatio* of the days and moments than to the *computatio* of pennies and pounds. Nevertheless, the Frenchman praised the progress made by the Greek and Arabic astronomers towards the *praecisa veritas* of accurate time-measurement, maintaining that the antiquated knowledge of Christian *compotistae* had to give way to them.

However, the modernist cardinal merely repeated the old proposals put forward by Grosseteste and Bacon. Using Bacon's words he conceded something that had ceased fully to apply ever since Alfonsine tables were composed, namely 'that the true length of the year is still not known to us with complete certainty'; like Reiner of Paderborn before him, he recommended the use of the ancient Hebrew calendar as a guide. How, then, did progress towards accuracy differ from a relapse into tradition? Since astronomy was still unable to produce any exact dates, conscientious council Fathers preferred to postpone the reform project. Science had apparently not progressed very far with measurements, and if Christendom was to trust reform, the reform had to be scientific.[169]

Science, however, was no longer the prerogative of Latin scholars and clergymen. In 1391 the most important English poet of the Middle Ages, Geoffrey Chaucer, wrote a treatise in English on the astrolabe and taught his son how to measure and calculate ecclesiastical bell-time, to *calcule*, using the *calculer*, the 'clock-hand' of the astrolabe. The layman, however, did not strive to achieve the same accuracy as an *Astrologien* who used Alfonsine tables; he also left the calculation of the 'holy daies in the Kalender' to specialists. All a tradesman had to do was find his way in time and space, on water and on land, number the current date in the Julian solar year precisely to the nearest day and hour, and to fix his position and the points of the compass by surveying the stars. To denote the measurement of short times, one of Chaucer's pupils borrowed the French noun *compte* from the *Romaunt of the Rose*.[170] In 1413 another follower of Chaucer, probably the Benedictine monk John Lydgate, Latinized the word into *compute*, meaning the computation carried out by the long-range calendar; Lydgate introduced the word *computacioun* c.1420. This did not mean that laymen were racking their brains over computation. Their working day centred on more immediate concerns and more important events.[171]

Nicholas of Cusa had in mind this indifference on the part of laymen when he wrote his *De correctione kalendarii* for the Basle Council in 1436. He was dispassionate in emphasizing that it had not hitherto been possible to measure precise shifts in time (*punctalis veritas*), even with the largest instruments. There was no hope of scientific progress; planetary movements and human understanding had no common measure. A discrepancy (*disproportio*) existed even between the paths of the celestial lights themselves; future regularities could not be inferred from earlier ones. From Alfonso the Wise onwards, astronomers had, in their own subtle way, been even more eager to attain accuracy than *computistae* like Sacrobosco, and even they had, in their crude way, *modo grosso*, forced the whole of universal time into too exact a frame, using fixed days for the spring equinoxes, and regular cycles for the revolutions of sun and moon.

In order to respond more swiftly to future fluctuations of time, the council had to cancel an entire week between Sunday and Monday at Whitsun 1439. 'Since it is a movable feast, the general public, *vulgus*, does not think about what particular day [of the month] it falls upon.' Moreover, the combined solar and lunar cycle of the Latin scholars was to be replaced by the pure Byzantine lunar cycle, and eventually the calendar year had, when necessary, to be shortened by a leap-day, initially every 304 years. Nicholas encountered two objections: first, that astronomers (*calculatores*) using Alfonsine tables to reckon would be confused, and secondly that economists who had agreed on deadlines and interest payments would suffer. Cusa expected interim solutions of both groups in view of a religious revival that would bring together Jews, Greeks and Latin-speakers, and would keep the Basle Council for ever in people's memories as the founder of a new era. For the first time since Augustus, a new calendar marked the beginning of a new age. If 'modern times' were heralded anywhere, then it was here. It failed to come to this because the council, already divided, wished to avoid creating any further occasion for discord and, no differently from ordinary people, had a fear of experiments with an open-ended future.[172]

In 1439 and 1474, two Viennese university teachers, Johannes of Gmunden and Johannes Regiomontanus, designed Latin calendars that were immediately translated into German and soon appeared in print. They computed the current phases of the moon for half a century in advance, providing important chronological points that everyone could check. They did not, however, make any general proposals for reforming the calendar. They merely wanted to see its dates fixed for the foreseeable future.[173] The fifth Lateran Council of 1512–17 again postponed the reform of the calendar because astronomers were still unable to show the exact correlation between the solar year and the lunar month. These educated men had come to fix their hopes more on accurate time-measurement than on the coarse reckoning of time. At the end, as at the beginning, however, the epoch of the computus did not regard determining times as an end in itself. Medieval Europeans did not intend to persist indefinitely with the ancient calendar; nor did they want to start afresh in a modern future. They simply wanted to make their present world bearable for the time being.

12

The Universal Machine and Chronology in the Early Modern Period

The age of perfection began with Canon Nicholas Copernicus. In 1543 he reminded Pope Paul III of the last Lateran Council and his *questio de emendando kalendario ecclesiastico*. In doing so, he justified his 'more precise computation of times [*supputatio temporum*], required to work out the movements of the planets [*in motibus caelestibus calculandis*]'. However, Copernicus did not rely solely on arithmetical calculations. He also used astronomical measurements like Hermann the Lame, the only difference being that he worked with more modern instruments than the astrolabe. He used these to compare the chronologies of the ancient Egyptians and Greeks rather than the cycles of Christian historians, making no mention of the *computus ecclesiasticus*. The newly discovered regularity of the planetary movements in God's *machina mundi* surpassed all the medieval chronologists' conjectures, as well as the hypotheses of Oresme and Cusa. Science had progressed. Since people now had an overview of the long period of their development, they were at last able to see the whole truth. Mathematics supplied them with laws that uniformly obtained everywhere in the open universe, in heaven as well as on earth, promising to make them into masters of their world and their time.[174]

The fact that people were not yet in any hurry to acquire these insights is most graphically documented by the farmers' almanacs of the sixteenth century. They provided their purchasers, most of whom could scarcely read or count, with a

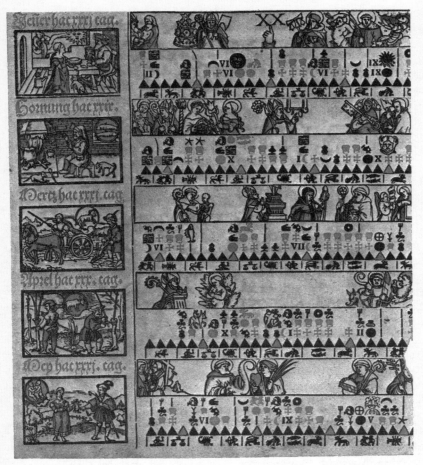

Plate 17 *Farmer's calendar for the leap year 1544. Broadsheet by the town printer, Christopher Froschauer, Zurich, 1544. Easter Sunday, 13 April, is the thirteenth triangle in the fourth row, below the flag with the cross.*

whole range of graphic symbols for lucky days and days of disaster, for wind and rain, the times for blood-letting and haircutting, warnings of wind and snow, the number of hours of sunshine, the lunar phases and signs of the zodiac, Sundays and saints' feasts. They reflected an entire cosmos of deep-rooted time-regulation, a mixture of faith and superstition, experience and prejudice, that could not be called into doubt by any science of mathematics. When the Reformation challenged the Catholic cult of the saints, even its pioneers did not think of making changes to the calendar.[175]

From 1559, however, Protestant historians established a new form of great year: the century. In a multi-volume collective enterprise, the so-called Magdeburg Centuriators set out to show how the papal church had falsified the Gospel of Jesus Christ; its editors assigned regular, clear intervals to each author, a hundred years per volume. They were clearly not intending to emulate the custom of the Catholic jubilee year used by Boniface VIII to commemorate Christ's birth; and they scarcely had in mind the *annorum centenarii* of Alexander von Roes, the end of which was supposed to herald the world's demise at that very time, in the sixteenth century. The new chronology of centuries was initially no more than a technical stop-gap, but one that soon made historians more inclined to arrange events and documentary evidence in chronological order, even when, as in the majority of cases, no firm date was known. The Centuriators' concept of time was no closer to that of Copernicus than Herodotus's was to that of Plato.[176]

The Catholic Council of Trent was irritated by the advances made in mathematics. It once again impressed upon all priests the need to learn the *computus ecclesiasticus*. However, exactly what *computus* they had to learn, Durand's vulgar one or Copernicus's astronomical one, was something even the council Fathers themselves did not know. It was not until the arrival of reform-minded popes that the council was relieved of the decision. Even the popes hesitated, until the Copernican belief in the regular operation of the universal machine spread, almost to the end of the third Easter cycle formerly computed by Abbo of Fleury. In February 1582, Pope Gregory XIII set about the Gregorian reform of the calendar, suppressing ten days in 1582, fixing the beginning of spring on 21 March, re-regulating the omission of leap-days, and determining the structure of our calendar to the present day. The reform of the calendar introduced a greater level of precision, but at the price of reduced consistency. The pope combined the reform with a new version of the saints' catalogue and the missal now incorporating instructions on time-reckoning. Only a handful of Catholics consulted these charts and the list of saints' feasts in the *Missale Romanum*; most of them consulted their pocket diary like everyone else. Even European princes applied religio-political rather than scholarly criteria when deciding either for or against the new calendar. The Christian denominations now even quarrelled about the dates on which they should celebrate the birth and Resurrection of their common Saviour.[177]

In a satire of 1585, the Italian heretic Giordano Bruno, an admirer of Copernicus, celebrated the pagan god Mercury as the Oracle of Mathematics and *compotista mirabile*; but in the infinite worlds of the myth imagined by Bruno, more fantastic and irrational symbols prevailed than those of school mathematics that he held in contempt.[178] On earth, the plain yearnumbers were so remote from ordinary life that in about 1585 the French nobleman Michel de Montaigne protested against the reform. His rural neighbours divided up their days as before while he himself lived in 'the years when we count differently', when 'no one has the time any longer to become someone else'. For him there was no other time-reckoning, *compte du temps*, than the solar year; it was as ancient as it was inaccurate. What, then, was the point in reckoning and improving? Modern French found no use for the word 'computist'.[179]

The immediate future belonged to a new science acquiring the name of *chronologia*, a humanist artificial term first coined in the sixteenth century. Its founder, an exile to Switzerland and the Netherlands, was the French Calvinist Joseph Justus Scaliger, the most famous scholar of his time. First, in 1583, he edited and annotated ten instruction books on calendrical computation for various nations and times: Hebrew, Ethiopian, Coptic, Syrian, Arabic, Greek, Armenian and Latin. He called them all *computi annales*, although he knew the word *computus* was attested late, and not before Firmicus Maternus. Scaliger also gave the name *computus* to his revised and condensed version, at least in its chapter headings, interpreting it as a *doctrina annalis*, a doctrine of years. Its medieval title once again held out the prospect that the summation of all *computi* would lead to one timeless truth concealed behind the many opinions voiced at different times.[180]

Scaliger himself relinquished this hope in 1606, in his final and most important work, *Thesaurus temporum*. First, he reconstructed and edited the oldest handbook on Christian time-reckoning, the chronicle of Eusebius and Jerome, complete with sequels. He then gave prominence to the most modern calendar, the Arabic; using mnemonic verses, he summarized its principal rules in a *computus manualis* so that travellers and merchants in Turkey would know how to use it. However, neither the moderns nor the ancients had full possession of the truth. The remaining objective was no longer to determine the divine or natural order of times, the beginning and end of time itself, but rather the historical specification of factual events; this was to be done with far more accuracy than had been attempted by the Magdeburg Centuriators with their vague enumeration of centuries. Since Scaliger conceived time as the 'range of celestial motion', modern *chronologia* must in his view be based on progress made by astronomers. Nevertheless, he regarded the Alfonsine tables as more accurate than the Copernican; newest was not necessarily best.

On its own, reckoning failed to attain the historical objective; even measuring was of no more than momentary help. As well as the natural sciences, those human sciences that preserved memories had to become involved. Using a critical, philological approach, Scaliger constructed a framework of fixed times from events recorded by the most ancient historiographers consisting of single days rather than sequences of years. These days fixed important points in time that helped posterity to date events: the destruction of Troy and the beginning of the Greek Olympiad calendar; the foundation of Rome and the start of the cycle of the Roman calendar; the biblical day of Creation and the day of Christ's birth; Mohammed's flight from Mecca and the era of the Seljuks. Instead of enumerating natural cycles, Scaliger's chronology noted high points of human endeavour that later became landmarks for historical thought. In short, it constituted historical time.

The earliest point in time was the least certain. When he tried to date the Creation, Scaliger arrived at the year 3949 BC, almost exactly the same figure as Bede. However, he did not share Bede's ambition to establish the true origin of world history; instead, his sole intention was to convert all foreseeable epochs into one another. To create some leeway for this, Scaliger assumed a period of 7,980 years for the entire *tempus historicum*, the product of the three most common yearly cycles in antiquity: 28 for the motion of the sun, 19 for the lunar orbit, and 15 for the indiction. He fixed the beginning of this period on 1 January 4713 BC, long before the advent of biblical time. When credible ancient Egyptian accounts forced him further back into no man's land, he added a *tempus prolepticon*, another arithmetical unit of 7,980 years, *more mathematicorum*.

By creating the space for a 'prehistory' *in infinitum* and for a future of at least equal duration, Scaliger separated chronology from the absoluteness and originality of all religious creeds, and bound it up with the relativity and progress of two technical procedures, the astronomical measurement of time and a critical-philological analysis of sources. As a Calvinist he criticized the Gregorian reform of the calendar because in his view it did not go far enough. He discovered no historical truths in the medieval *computus ecclesiasticus*, only 'the dreams of the old *computatores*, unequalled in foolishness'.[181] In order to save the Christian calendar, Scaliger's opponents resorted to operating with years 'before the birth of Christ', and this numbering, already used by Bede, became established not because it emphasized the first Christmas as the focal point in the history of salvation, but because it circumvented the uncertain date of the Creation.[182]

Slide-rules and calculating machines were already being made in the seventeenth century, the age of manufacture. Clocks and machines had been combined from as early as the fourteenth century, when model cockerels, crowing and with flapping wings, were activated by planetary clocks. But was it not now possible for machines to compute time as well, along the lines of Copernicus and Scaliger? The first modern calculating machine, designed in 1623–4 by the Tübingen orientalist and mathematician Wilhelm Schickard, was actually meant to assist the chronological and astronomical work of Johannes Kepler. What his friend was undertaking *logistice*, he was attempting to do *mechanice*, Schickard wrote in a letter to him. His *arithmeticum organum* was poorly suited to the purpose, since it was unable to add up numbers of more than six figures automatically, *datos numeros statim automatos computet*.

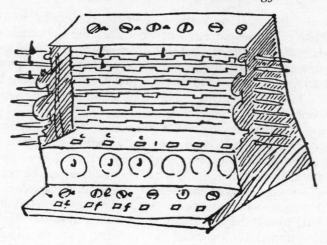

Plate 18 *Schickard's calculating machine, sketched by him in a letter to Kepler, Tübingen, 25 February 1624, now in the Kepler Bequest, Pulkowo-St Petersburg Observatory. The machine consisted of three separate sections. In the centre the main section, the calculating mechanism for adding and substracting, includes the cogs (ddd) and their meters (ccc). Of the cylinders in the upper adjustment mechanism used for multiplication and division, the knobs (aaa) and slides (bbb) are visible. The dials of the counting mechanism at the bottom are recognizable by the knobs (eee) and their meters (fff).*

Computare, however, signified all four basic methods of arithmetic. Technology was made a function of theory.[183]

In 1642–5 the young Blaise Pascal, son of a tax collector, invented and built a *machine d'arithmétique* for practical use. He meant to relieve the *calculateur*, and the tax official in particular, of the soul-destroying task of setting out numerous counting beads (*jetons*) or writing out long series of numbers in order to settle funds. The counters on the adding machine also corresponded to the currency system of the time. Although Pascal compared it to a pocket watch (*montre*), with one clock-maker (*horloger*) later using it as a model, Pascal did not recommend its use to time-reckoners. Why not? Following his conversion in 1654, Pascal used Augustinian incisiveness to distinguish healthy common sense and mathematical method from natural spirit and religious contemplation. We know how to use time and numbers correctly in our daily lives; but their Aristotelian definition turns our heads. Human beings alternate between the infinitely small and the infinitely large, between the void and the universe. It is only in God's immensity that the extremes of times and numbers truly come together; all we glimpse in these abysses are probabilities. For four thousand years the Jews have affirmed that they are the most ancient nation of all, and who are we to disbelieve them; but the Creation and the history of salvation are articles of faith, not sums.

Pascal challenged the rationalist mentality of his century, a century in which man was conceived, under Descartes's influence, as a mechanism comprising body and soul, as something calculable. In his view, each human being, having been created in God's image, was an intellectual being consisting of reason, and was merely *automate* by virtue of habit. What distinguished human beings from other living things which behaved completely instinctively was the thinking soul. On whose side, then, did a calculating machine like Pascal's stand? 'The *machine d'arithmétique* achieves effects that are closer to thought processes than anything done by animals.' The calculating machine seemed to symbolize human ingenuity in the midst of absolutist constraints; but artificial intelligence lacked even the simplest animal will. Consequently, the gulf separating living things from instruments was maintained.[184]

The many Europeans inventing and experimenting with calculating machines in the seventeenth and eighteenth centuries followed Leibniz's example in laying the main emphasis on mathematical theory, and merely wanted to free the minds of prominent thinkers from routine mechanical work. For this reason they never called their technical products 'thought-machines', but instead gave them the Latin name *machina arithmetica*; the French name, *machine d'arithmétique*; the German name, *Rechnungs-Maschine*, later *Rechenmaschine*; and the English name, *calculating machine*. Some were intended for astronomical use, but none were made for the purposes of computation. About 1660, the German Jesuit Kaspar Schott incorporated slide-rules into his *organum mathematicum* for determining Easter, but unlike Schickard, he no longer enlisted the terminology of *computus*.[185]

By the seventeenth century, those continuing to deal with time-reckoning in the old, computistical sense were almost

60 **Deß Abenteurlichen Simplicissimi**

Agricola Bischoff zu Cabilonæ. Wittburg Jungfr. in England. Eugenius, Pamphilianus, Castor und Serenus martyrer zu Nicomedia. Collegius Diacon, wie auch Rogatus und Satyrus martyrer zu Alexandria.

G ℞ Der 18. Mertz.

Alexander Bischoff und Martyrer zu Jerusalem. Gabriel ErtzEngel. Cyrillus Bisch. zu Jerusalem. Anshelmus Bisch. zu Mandua. Salvator, Franciscanus. Eduardus König und Martyr. in England. Narcissus Bisch. und Martyrer zu Augusta.

A Der 19. Mertz.

Joseph Pfleger Christi Mariæ Gemahl. XIII. Johannes Abbt und Beichtiger. Calocerus Martyrer. Apolonius / Bassus / Sorentus und Leontius. Theodorus Bisch zu Cæsarea. Lactinus Bisch. in Irland. Maria Magdalenæ Erhebung von Aquitania in das Vercelliacensische Closter. Menignus Walcker und martyr. Amandus Diacon zu Gent. Landoaldus Priester daselbsten. Sybillina Jungfraw Dominicaner Ordens.

B Der 20. Mertz.

Archyppus S. Pauli Gesell. Cut.

XV. Calendas Aprilis.

Diß ist der erste Tag der Welt / an welchem GOtt Himmel und Erden erschaffen hat; Nemblich uff einen Sontag.

An diesem Tag hat der HErr JEsus zween Blinde erleuchtet. Matth. 20.

Anno 1502. erhub sich der Bundschuch oder Bawren-Krieg umb Speyr und Bruchsall; Mit einm Wort Ihr Meinung war selbst Herrn zusein / aber sie wurden zertrendt / hin und wider uff mancherley Weeg hingerichtet.

XIIII. Calendas Aprilis.

Diesen Tag hat GOtt das Firmament erschaffen am einem Montag.

Maria Magdalena hat diesen Tag den HErrn JEsum zu Betania gesalbet. Joan. 12.

Ist auch auff diesen Tag geschehen die Vermählung Josephs mit Maria der ewigen Jungfrawen und Gebärerin Gottes; an welchen Tag er Joseph auch auß dieser Welt geschieden.

Eysen und Staal zuhärten.

Nimb Regenwürmb / Senff / Saamen und Rettig-Safft thue es untereinander / laß beym Fewr ein wenig erwallen / mach Staal oder Eysen glühend / stosse es hinein / so bekombt es ein solche Härtung / daß du ander Eysen mit schneyden und bohren kanst.

XIII. Calendas Aprilis.

An diesem Tag hat GOtt die Wasser gesamblet, Item

Simplicissimus.

Lieber Pfetter ich habs so wollen haben / laß der guten Mutter auch ein wenig Ruhe.

Knan.

Ja ihr werdet grossen Nutzen darvon haben / wann Regen einfiele und uns das öhmbt verdürbe; man muß drauff trucken wann die Sonn scheinet / weil alsdann ein Narr mehr Futter dörren kan / als sonst zehen Docter wanns regnet: Es heist was vor Michaëlis nit geöhmbtet werde / das müsse man hernach öhmbtlen / daß ist / daß öhmbt bettelhafftig und sehr langsamb einmachen / unnd wird doch nichts guths drauß / sonder keinützig-wetterfarbig roth Ding / dz nachgehents wann mans im Winter füttern muß / den Kühen nit wenig an der nachgeher / und nit wenig an der Milch schadet / wann sie anderster nit gar kranck darvon werden.

Simplicius.

Es wird drumb eben nicht so gleich regnen.

Knan.

Was? ich will mein Kopff verwetten wo es noch 3. Tag schön bleibt / dann ich habs ja gester wol am Heyland (so nennen die Bauren uff den Schwartz-Walt und im Preyßgaw den Mon / wann sie ihn ehrerbietig nennen wollen) gesehen: gehe nur hin Mutter zum Gesind waß du gleich nichts thust / als zusehen / so werden doch sie desto fleissiger seyn: indessen will ich mit dem Knecht in Walt / und ein baar Plöcher mit den alten Ochsen herab schleiffen.

Sim-

Plate 19 *Grimmelshausen's dating of the Creation. The perpetual calendar, Nuremberg, 1670.*

exclusively creators of vernacular calendars. From then on, time-reckoners were called in German *Kalendermacher* (calendar-makers). We might single out the most famous, the catholic poet Jakob Christoffel von Grimmelshausen. In his

Perpetual Calendar of 1670, the first of six columns assigned to each of the German-named days consisted of a martyrology listing the names of almost six thousand saints. The second column gave an account, using Latin dating, of all kinds of histories, the history of salvation and universal history for the common man. Under 18 March it read: '*XV Calendas Aprilis*: This is the first day of the world on which God created Heaven and Earth; that is, on a Sunday.' Under the same day it read: 'In the year 1502 the Bundschuh or peasants' war rose up around Speyer and Bruchsall.' Sublime, primeval time and present privation converged. The third column was devoted to the moment, combining 'farmers' rules for the day's weather and stories as brief as enumerated time: *Kalendergeschichten* (calendar stories).

In the fourth column a scholarly dialogue 'on calendar-making and related issues' ran through almost the entire year; it was a proper computus, only Grimmelshausen no longer gave it this name. Like Aristotle (in truth more like Plato), the calendar-maker declared that time was 'a number or expanse of the upper corporis [*sic*] of heaven'. He brought together the views of Jews and Christians who held that God created the world some time between 3707 and 6984 BC. He discussed at length the Christian computation of Easter, 'when calendar-making is at its most important'. He even supplied a brief universal history of the calendar and its reforms. The final two columns dealt consecutively with astrology and soothsaying. The Carolingian triad of martyrology, chronology and computus was thus revived, albeit as fallen cultural assets, in the notorious society of weather-makers, star-gazers and sooth-sayers. It was to them that the saying 'you're lying like a calendar-maker' referred in the writings of Grimmelshausen himself.[186]

In England in 1646 the sceptical physician Sir Thomas Browne adopted Montaigne's style, demanding that people should not attempt to achieve the 'exact compute of time' but should instead be content with the 'common and usual account'. Believing that they could illuminate the impenetrable darkness of time's origin as well as the *calculation* of ancients like Bede, and the *chronology* of moderns like Scaliger, self-opinionated scholars had a deterrent effect. Even more irksome were the popular assumptions, 'the calendars of these Computers', the weather-rules of calendar-makers who even failed to notice that the Gregorian calendar had only been introduced in Romance countries, while Great Britain and parts of Germany retained the Julian Calendar. These *computists* challenged the usual account by which the year numbered 365 days.[187] It would take another century before the English could bring themselves to introduce the new calendar. In modern English the words *computer* and *computist* sounded like the Middle English *compute* and the Middle Latin *computista*. Browne, however, deprived the computists of what little remained of the reputation which Bede had gained for them nine hundred years earlier.

The word *computer* made another appearance in 1707, in a satire by Jonathan Swift. But instead of attacking medieval computation, he reached far ahead of his time to attack modern computer science, ridiculing the modern scholars who, in contrast to the ancients, would read and think nothing, but instead just collect everything. They study books from their back cover, merely leafing through abstracts and indices, using these to create many other books, even though all the really new ideas could quite easily be included in a single volume. Swift claimed that a 'very skilful computer' had told him so; he had proved it by arithmetical rules. In truth, this *computer*, a theologian and believer in progress, knew nothing of the science of number or time-reckoning, producing large numbers of books and profiting from the prevailing fashion.[188]

In 1726 Swift described a huge machine owned by an 'advancer of speculative learning', allegedly constructed using strict *computation*, and operated by forty people. With the help of this machine, 'the most ignorant person at a reasonable charge, and with a little bodily labour, may write books on philosophy, poetry, politics, law, mathematics and theology, without the least assistance from genius or study.' If the public were to build five hundred such machines, the world would soon possess 'a complete body of all arts and sciences'. This device, storing all words in the language and reassembling them in ever new ways, would today be described as a computer used for processing non-numerical data. That this writing machine would be given the same name as the writer

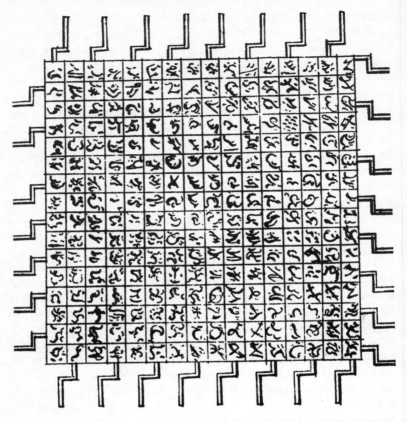

Plate 20 *Swift's writing machine for advancers of speculative learning, a copperplate engraving designed by Swift himself in the first edition of* Gulliver's Travels, London, 1726. *The diagram is a simplification, for 256 dice, each with six sides, are not sufficient to record all the words of a language; also, only 31 of the 40 cranks are visible. The script is a mixture of Hebrew, Arabic and Chinese.*

himself was something not even the waspish Swift could have suspected at the time.[189] His satire nevertheless attacked the fundamental principle of the calculating machine, associating it with the term *computer*. The time used by the device and its operator was the ahistorical moment rather than the universal history, its number the amassed quantity rather than qualitative evaluation, its language a system of symbols lacking any deeper meaning.

13

Chronometry and Industrialization in the Eighteenth and Nineteenth Centuries

The eighteenth century saw the beginning of changes in the European concept of time. These were too uneven, too slow in forming, to be described as part of a revolution, but they finally brought the 1,400-year-old history of the computus to an end. Two opposing developments in technical chronometry and historical chronology combined. In the course of the century French and English clock-makers succeeded in constructing precision clocks which not only indicated the seconds on their dials, but also kept time accurately to the second, earning them their new name: in French, *chronomètre* from 1701, in English *timekeeper* from 1686, and *chronometer* from 1735. It was the chronometers which first realized the demand for time-measurement stimulated by the mechanical clock in the fourteenth century, bestowing global mobility upon Europeans in the age of expeditions. The navy, including the merchant fleet, acquired instruments for determining their exact position and for deliberately changing course on the high seas.[190] The perfection gained in local time was lost in universal time. The interest of scholars in grouping together years in chronological sequence waned because it was not possible to pinpoint one particular beginning for the complex history of the world. Clearly, no ancient time-scale extended back to the real beginning; the Jewish-Christian calendar, already called into doubt by Scaliger, did not bear comparison with reports from Egypt and China.

Both developments were compounded by a philosophical transformation that revived the Platonic concept of time. In 1686 Isaac Newton divided time even more radically than Abbo of Fleury had done at the turn of the millennium; he differentiated between mathematical or chronometric time on the one hand, and historical or chronological time on the other:

> Absolute, true and mathematical time flows into itself, and by its own nature, equably, without reference to anything external; it is otherwise given the name duration. Relative, visible and ordinary time is a perceptible and external measurement of duration by means of motion, be it exact or inequable, which one ordinarily makes use of in place of true time, for example the hour, the day, the month, the year.

Scholarship sided with truth rather than with custom.

In 1703 Gottfried Wilhelm Leibniz challenged Newton, refusing to separate time, something merely relative, from events occurring in succession, particularly the phases in an individual's life. Nevertheless, he also dismissed Aristotelian theory, conceiving both historical and mathematical time as a continuum: 'The present is pregnant with the future and laden with the past.' The succession of human ideas, however, never corresponded to 'the flow of time which is an equable and straightforward continuum, like a straight line'. Like Aristotle, we could regard time as a measure of motion; we could see equable motion as a measure of inequable motion, and view duration in terms of the number of periodic movements, for example a certain number of orbits of the earth or the stars. As with Plato's harmony of the spheres, however, nothing permanent would remain of this if, as Nicholas of Cusa had suspected, the motions of heavenly bodies were subject to temporal changes. This was doubtless true of the daily orbits of the sun, and perhaps of the annual ones too.

It is, then, not particularly sensible to follow Scaliger in presupposing that years remained a constant length in universal history. Julian chronology, whose preliminary period included the possible, was certainly more efficient than biblical chronology, which lacked the capacity to add together the years lapsed since the world's beginning in a time sequence. No one can reduce with impunity the long evolution of the world, during which nature developed at a steady rate, to one year of creation. It is almost as presumptuous to ask how long one day and one night would have lasted in Scaliger's pre-history 'before there were days, nights and years designated by an orbit of the sun'. Despite all the progress that has been made, determining time in the present is subject to differences which must have always been there, but which we would not discover until some point in the future. 'The rotation of the earth on its axis . . . is our best measure to date, and tower-clocks and pocket-watches (*les horloges et montres*) help us divide it. In the rhythms of time, however, this daily rotation of the earth can also change.' A continuum such as time can be founded neither on periodicity nor on numbers; they resemble each other much less than times. If the Aristotelian association between time and number is dissolved, historical time also has to be divided according to non-chronological rules.[191]

From 1725, in a *tavola cronologica*, Giambattista Vico tried to salvage the older origins of the world represented in the Jewish calendar and to prove that the 4,000 pre-Christian *anni del mondo* depicted in the Old Testament were objectively accurate. However, he went on to explain that, during the most ancient eras, people counted in harvest cycles rather than years. As the originators of time-sequences, the earliest myths depicted gods such as Saturn-Kronos-Chronos, the god of sowing and the god of time, or heroes like Hercules who created the space for farming by fire-clearance, and devised the Olympics. If astronomy and mathematics, and consequently rational time-measurement and time-reckoning, made their first appearance among the Chaldaeans a thousand years after the Flood, the Bible offered no data for measuring the mythical time that went before. The imaginative *cronologia poetica* of different nations could be arranged in a typological series, even in a cycle, but not in a numerical series. If Scaliger's modern chronology was caught up in mythology, then this was even more true of Bede's medieval computation.[192]

The most we can surmise is that some nations practised perfect time-reckoning at an earlier stage than others. In 1756 Voltaire believed the same thing of the ancient Chinese astronomers as Vico did of the ancient Jewish patriarchs. 'They invented *un cycle, un comput*, beginning 2,602 years before our

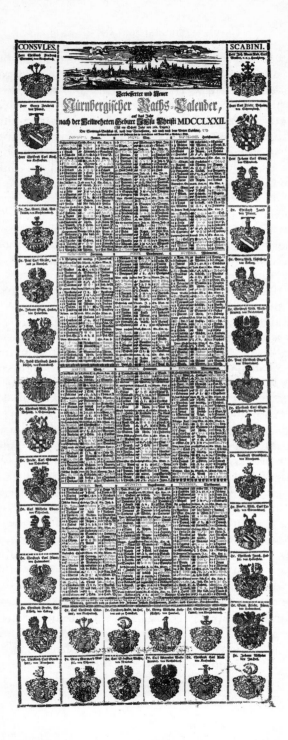

own. Have we the right to challenge their unanimously accepted chronology? We have sixty different systems *pour compter les temps anciens,* and thus have none.' Voltaire satirized Christian *chronologistes* who defended the biblical approach, and he derided the chronology based on Christ's birth as *notre ère vulgaire*. It was the invention of imitators and ignorant people, as Newton's discoveries relating to celestial mechanics conclusively proved. But Voltaire did not know how old world history was, nor did he wish to; instead he wanted to 'write the universal prehistory of the French bourgeoisie'.[193]

Germans translated *chronologia* as *Zeitrechnung* (time-reckoning), a word familiar to Kepler and widely used from 1716, but they interpreted prehistory more mythically than the Italians and the French. In 1771–2, Johann Gottfried Herder brooded over the first book of the Bible: 'What an arduous task time-reckoning is. How difficult is the abstraction of a series of numbers? How much more difficult is remarking the months, seasons and years by a recurrent sequence of days.' Any *chronologie* that failed to perceive God as the Creator of time or accept the existence of a celestial time-reckoner, a calendar in the clouds, did not extend back to the dim and distant origins of humanity. Among the hieroglyphs symbolizing what was to come there was not one numerical symbol. People first learned how to reckon time within separate nations; Greeks did it differently from Romans, Jews differently from Christians. Christian dating was only one possible method among many. In the age of incipient historicism people spoke with increasing frequency about the 'first century in our calendar' and of centuries 'before our calendar', referring less and less frequently to years before or after the Incarnation of Jesus Christ.[194]

While the origins of the world were buried in myth, measurable time was orientated towards developments up to the present. In 1793 the French revolutionary calendar dismissed all religious chronologies as irrational, especially the Christian

Left: Nuremberg Council calendar for the leap year 1772, printed by the council printer Johann Joseph Fleischmann, Nuremberg, 1772. It includes the shields of the councillors, Sunday letters, Saints' days, signs of the zodiac and weather forecasts.

Plate 22 *The French revolutionary calendar, a copper engraving by Louis-Philibert Debucourt, Paris, 1794, now in the Louvre, Paris. Revolution is portrayed wearing a cap of liberty, reading and dictating from a book of astronomy; at her feet are a simple sundial and ancient documents, one bearing the title 'Calendrier Grégorien'.*

ère vulgaire, and showed the free, equal and fraternal time of all humanity as beginning on 22 September 1792. The metrical system for measuring capacity was eventually introduced world-wide. The corresponding plan for measures of time was more readily introduced by authorities than private individuals, and was implemented more quickly in the town than in the country. It failed because of the local, Parisian year which began at the autumn equinox, the affected ancient Roman names of the days and months, and the abstract decimal division of the days and hours. Nevertheless it bequeathed to the French study of history an enduring respect for chronology as the arrangement of permanently memorable things, together with a regard for *histoire contemporaine* and its ever-increasing and often surprising events in which lasting things were likewise manifested. 'The event resembles a stone thrown into water; it creates ripples and stirs up mud from the bed which brings to the surface things that creep and fly under the bedrock of life.'[195]

The associations sought by nineteenth-century German historicism, itself a creative response to the challenge of the French Revolution, were not found in times of festivity or upheaval, nor in present events vaguely called *Zeitgeschichte* ('contemporary history'). In 1854 Leopold von Ranke preferred not to define epochs in political history as sudden and decisive points in time; for him they spanned at least a generation, usually a century. In 1868 Jacob Burckhardt relegated chronology to the level of a positivist *instrument*, while cultural history – 'the clock which tells the hour' – amasses enduring events from all periods. 'Historical time is not simply measured time. It is time that has been lived through, suffered, and experienced. It is determined not by the hand of the clock moving forwards minute by minute, but by the far more arhythmical clock of internal and external experiences.' Exceptions such as Theodor Mommsen and Bruno Krusch prove the rule, namely the low regard for time-reckoning among German historians that I described at the outset. They abandoned social time, leaving it to journalists and antiquarians.[196]

Non-scholars, for their part, were taking short spans of time more and more seriously, beginning with leisure time, the older form of which was regarded as a time-wasting exercise. After preliminary steps which, in the case of English horse-racing, began as early as the seventeenth century, early nineteenth-century Europeans encouraged vigorous and purposeful physical exercises, full of temporal dynamics and selective tensions; they aimed for an increase in performance, measuring records in seconds. Modelled on the marine chronometer, the small stop-watches needed for this purpose began to be developed in 1825; from 1850 their operators were

Plate 23 *Pocket watch, Saxony, c.1900, with roman numerals for the hours, arabic for the seconds.*

named after the measuring instruments themselves: *time-keepers*, *chronométreurs*, *Zeitnehmer*, *cronometristi*. Ever since, popular sport has been unthinkable without temporal standardization and numerical mechanization.[197] The fact that daily newspapers, railways and the telegraph accelerated communication over long distances transformed the public sector in a similar way.

In the late nineteenth century similar endeavours were made in the realm of work, beginning with industry. The loom, the blast furnace and the steam engine did not spare people any expenditure of labour as did older machines. They consumed people's time and converted it into merchandise. When entrepreneurs replaced hourly rates of pay with piece-work in order to increase efficiency, British foremen were given the name 'timekeepers', because they ensured that hours of work were strictly adhered to. In the 1880s the American engineer Frederick W. Taylor began his *Time Studies*, approaching the worker as a living machine. The slightest movements by the workforce were measured using stop-watches, labour costs were calculated, and norms were set for the ratio between time and pay. 'Timekeepers', 'Zeitnehmer' and 'Zeitrechner' were the names now given to the people who did the monitoring, as well as to the instruments they used to do it. Time became a calculated standard value, enforcing perfect operation and excluding the unexpected.[198]

The difference between human beings and their instruments disappeared when saved time was valued more highly than given time. In literature, this automation of the present that made workers into robots stimulated the beginnings of 'science fiction'. H. G. Wells's story *The Time Machine* of 1895 became a model for the genre. The researcher, for whom the grandfather clock in the workshop indicates every passing minute, is transported by a technical instrument to the year 802701 AD and back again, all within twelve hours. The clock on the time machine indicating this date to him is a calculating machine that works on Pascal's principle. Although it follows the alternation of natural days, its four dials jump from one power to another: single days, a thousand days, millions of days, a thousand million days. Torn from the continuum of his limited time, the man returns injured from a future still suffering from social envy and machine mania, and heralds the end of the world. If narrated time is the same as enumerated time, nothing is to be gained from mechanically accelerating time and amassing numbers.[199]

In the nineteenth century, described by Edmund Burke disparagingly as the age of the *calculators*, instruments for reckoning were still frequently called *calculating machines*. However, they were no longer being made as single, hand-crafted models. Instead they were now manufactured as industrial, mass-produced commodities, given attractive names, and sold in all the industrial nations. In commerce, industry and administration they rationalized the rapidly increasing number of arithmetical tasks that required little intelligence

but were costly in terms of time and labour. In 1820, for instance, a French *arithmomètre* began performing calculations
for insurance payments and technical designs. From 1885 an
American counting machine called the *comptometer* proved to
be an efficient office machine for listing sums of money and
products. In 1884 the American statistician Herman Hollerith
began constructing an electric tabulation system with clock-
style counters that was later used to simplify the United States
population census of 1890. Particularly in North America,
technologists bestowed new honours on the word *computus*,
reserving the terms *to compute, computation* and *computer* for
people and their purer mathematics. Even Pascal's philosophy
and Swift's satire probably continued to have an effect.[200]

   In 1897 a British engineering journal crossed this dividing
line, giving the name *computer* for the first time to a newly
invented calculating machine that functioned 'in the manner
of a circular slide-rule'. In the same year, coincidentally, astronomers suggested to Pope Leo XIII that by introducing a
perpetual calendar with a fixed Easter date, it would be possible to dispense with computation altogether.[201] A few years
later Hollerith's punch-card machine was also called a *statistical computer*, and in 1911 his firm was renamed the Computing Tabulating Recording Company. Industry has since settled
upon the old terminology.[202]

   In 1927, the pathos of industrial progress was dismissed as
common existential oblivion by the archaic philosophy of
Martin Heidegger. It revealed the original temporality of
human existence, whose concerns are directed towards the
future and death, and dismissed the arithmetical 'fixing of
time' as an inner temporal intervention in a supposedly endless present. Nevertheless, Heidegger not only implied that his
contemporaries possessed this vulgar concept of time, he also
attributed it to the whole of past history, to the time-theories
of Aristotle to Scaliger as well as to practical time-reckoning
and time-measurement. He described the ancient peasant
clock, which measured lengths of shadow cast by the sun
relative to the lengths of the human body and foot, as if the
only way this 'timepiece' differed from the hands and numerals on the modern pocket watch was in being less accurate and
less public. However, Heidegger did not find any contempo-

Plate 24 *Newspaper advertisement for punch-card clocks, Berlin, c.1920.*

rary historian claiming that the long centuries of the water-clock and hour-glass had not truly objectivized their borrowed time, and not one modern sociologist said to him that it was only in the last decades of the stop-watch and the punch-card clock that their invalidity was frantically suppressed.[203] The wholesale denial of the human sciences brought to an end what the enthusiasm of the natural sciences had begun: in the history of words, the computer has killed off the computus.

14

Computers and Atomic Time in the Twentieth Century

What about the history of objects? Between 1937 and 1946, North Americans invented various prototypes for electronic data processing machines. The participating mathematicians and computer scientists, some of them working in collaboration with Hollerith's successor company IBM (International Business Machines), abbreviated the usual term *calculating machine* to *calculator*, and from 1940 also began abbreviating the analogously formed *computing machine* to *computer*. They were even more given to devising practical acronyms such as ABC (Atanasoff-Berry Computer), and ENIAC (Electronic Numerical Integrator and Computer). They dispensed altogether with the distinction between, on the one hand, the man operating the machine, and on the other the instrument performing even the most complex calculations with greater speed and accuracy than he could.[204] As though Pascal and Swift had never put pen to paper, the word *computer* has since been used in English almost exclusively to describe these machines. Most other world languages use the English word to describe them. Many French people, however, shunned this Anglicism, and in 1955 they coined a new, similarly personalizing yet more apposite word: *ordinateur*.[205]

These computers cease more and more to resemble Pascal's calculating machine. Even if some of the symbols they work with are interpreted as numbers, computers now utilize all kinds of symbols. Yet the inventors for their part compared their 'Rechner' ('computer') not only to the seventeenth-century machine, but also to the tenth-century abacus. Many of them referred to it because they believed in linear progress, and were seeking early stages of development for it. Even those who talk of the *digital computer* leave out two turning points in history. After Bede had described finger-reckoning as *computus vel loquela digitorum* in the eighth century, Gerbert of Aurillac went on to use the term *digiti* in the tenth century to denote the numerical symbols for 1 to 9, though he now transferred them as counters to columns of the abacus rather than counting them on ten fingers. The term was retained in its English form, *digits*, to denote single-figure numbers; today, however, *digital data* are no longer figures in the decimal system, but rather standard symbols assigned to each of the symbolized values by a binary code.[206]

Just as questionable is the fresh attempt, initially made in North America, to integrate the computer into the continuity of the medieval mechanical clock. Certainly both altered human perception of the world, thereby creating new realities. However, computers are fundamentally different from clocks. Far from creating an awareness of time, their capacity enables them apparently to process everything at once, thus rendering time insignificant. The symbols they use to express time are not 'analogue' symbols, hands moving on continuously 'round the clock', but variable 'digital' signals that on retrieval are illuminated in truncated lines. Just as dials no longer resemble the round earth, and their numbers no longer resemble ten fingers, the single point in time no longer corresponds to the moment, and the flow of time is no longer analogous to the life-history of human beings. Even the linkage of time to the movements of heavenly bodies, the sidereal hour, the earth day, the lunar month and the solar year, was removed when the perfection of the new clocks exposed the fluctuations in the earth's rotation assumed by Leibniz. Since 1974 the second has been measured in atomic vibrations, and each second is subdivided into a billion nano-seconds countable by computers. Are these, too, time-symbols designed by human beings to fit into the social context of their relations? We are no longer living in a calendar time; we are living in an atomic time.[207]

Historically, the computer can be conceived only as the

product of two modern developments, the first culminating in the seventeenth century, the second in the nineteenth: the mechanization of our world-view and the industrialization of skilled manual work.[208] As a 'time-machine', the computer does not project the researcher as far into man's future and past as modern 'science fiction' would wish. In 1973 it was possible for a computer to supply the times of the new and full moons of the previous two and a half thousand years within 132 seconds, a calculation that took Hermann the Lame months to work out on his abacus, and that Leibniz would like to have performed on his calculating machine. Historians today profit from the fact that the same long-range time sequences of astronomical chronology guide the spaceflights of our modern missiles. On the other hand, even antiquarian historians ignore a proposal made in 1968 that computers should be employed to produce a synopsis of ancient and medieval calendars and hence to resume the work begun by Scaliger. Historical chronology may still be popular among writers, but not so among scholars.[209]

The need for numerical data processing is greatest for precisely that era whose demands for infinitely fragmented time and infinitely increased number have led to the computer: the age of mechanization and industrialization, its econometry, demography and socio-mathematics, its 'price and salary curves, and the graphs of birth-rates'.[210] Nevertheless, if today's historians take the time to glance up from their special papers, they will not evaluate the computer solely according to the functions it either performs or does not perform for them. The computer has become a historically important symbol for the late twentieth century, a symbol of logical perspicacity in the midst of senseless information.

The ancient sundial and water-clock, the computus of the sixth century, the eighth-century bell, the astrolabe of the tenth century, the fourteenth-century mechanical clock, the seventeenth-century calculating machine: with all of these the computer shares the rationality of an instrument that helps humans to understand their world. Societies throughout history have always liked to stylize the latest 'state of the art' instrument into the epitome of their present. However, they have never enveloped earlier devices in the aura of a symbol intended to be a substitute for their world. It was neither a European historian nor a European social scientist, but an American computer expert who observed in 1976 that all of us, experts and laymen alike, 'have made the world ... into a computer.' 'This remaking of the world in the image of a computer' imperceptibly alters the concept of language and sign, and of time and number, for all fellow human beings.[211]

In 1987, for example, a German encyclopaedia stated that computers allow us access to millions of *data* in a matter of seconds, and now, a mere fifty years later, they have already reached their fifth *generation*.[212] It seemed to ignore the fact that until recently the words *datum* and *generation* had more meticulously defined a time-frame that was far wider! From the early Middle Ages onwards *datum* meant the laboriously enumerated day on which a decree was finally issued after long preparation. Following Jerome's translation of the Bible, *generatio* meant the thirty years of an entire human generation, or even the hundred years collectively attained by grandparents, parents and children.[213] A computer can, of course, store more data than are left by all the days in a human life; in no time at all it disposes of the menial tasks at which a whole century had previously laboured. However, it only grants more freedom to those who do not confer upon it the characteristics of a god to whose perfect rationality fallible and unpredictable human beings submit themselves. Its effect is too quantitative and momentary for it to become the measure of the world, humanity and time.

Prominent thinkers in the natural sciences have known this for a long time, and have warned against a breathless levelling-out process that would gear all the many different divisions of time to the requirements of the laboratory and the facilities of the computer. The Viennese computer scientist Heinz Zemanek pioneered a new computer in 1961, and since 1978 he has become absorbed in the future and history of time-reckoning. On the one hand, he holds out high hopes for the nano-second rhythm of a new Co-ordinated Universal Time that could soon synchronize several computers, and looks forward to research and technology advancing towards even smaller time-units than nano-seconds; on the other hand, he wishes that God would forbid any further reform of the calendar. (Austrian computer scientists planned a modern revival of the French revolutionary calendar with their '1984 Calendar'). In the age of the computer, the few benefits reform would bring could be achieved by other means. Its main effect would be to destroy the flexible order of all living creatures, and in particular disrupt the basic cycle of day and night, sun and moon, that the ancient computus had carefully preserved. 'Whatever time may look like tomorrow and the day after, it will need a temporal order that simultaneously satisfies the demands of accuracy and guarantees the best link with the past.'[214]

It is doubtful whether it still lies within the power of the physicists and the historians jointly to create a future temporal order that is as precise as it is stratified. In 1987, after thirty years of specialist research on all aspects of time, the American philosopher Julius T. Fraser presented a set of historical findings that gave little hope to those eager to press on, and to a large extent justified the fears of the sceptical. In the last generation, according to Fraser, the global networking of the present, starting with the atom bomb and the computer, has brought about increasingly dense collective and atomized concurrences. It has destroyed more and more of the complexity of past temporal orders, and narrowed the range of variations between the working day and free time, youth and old age, biological, intellectual and social development. Mankind is on the way to losing the scope of its humanity and becoming simply an anthill.[215]

15

Calculable and Allotted Time

If we consider not only the connections between man and nature, and between nature and society, but also the relationship of man to himself and his kin, then better prospects are opened up, and more important problems present themselves. As a humanities scholar I have to remember that the computer is predictable, but it is not accountable. The activity of measuring and calculating starts to involve more than just tinkering with numbers as soon as the amount of time allotted to man begins to have an effect. Standards have to be set from the beginning, commitments and distances have to be maintained along the way, and afterwards account must be given. As the present study set out to show, the human history of time and number since Plato and Aristotle has never centred purely on moments and quantities, it has always been a matter of duration and quality too. The living were repeatedly confronted by the old question of whether they should raise themselves above their momentary existence, come to terms with it, or lose themselves in it. Each epoch came up with many different responses to this question, all addressing the contradictory aspects of time.

Which of these responses made history was ultimately dependent not on the circumstances and expedients of the age, nor on whether it was geared to the past or the future, nor even on the symbolism it used to try to master its present. It depended instead on the sincerity and circumspection with

Plate 25 *'On losing time', woodcut by the Petrarca Master, Augsburg, 1520. The illustrated text (Francesco Petrarca, De remediis utriusque fortunae, II, 15), makes no mention of clocks, only the brevity of a lifetime. The hour-glass on the table has apparently run out. The hour-hands on the two mechanical clocks, the tower-clock on the right and the wall-clock on the left, are pointing to 12 o'clock; the fact that time has come to an end is also indicated by the zodiac sign of Pisces, the last of the year. The object behind the hour-glass on the wall is not identifiable (possibly the float of a water-clock?).*

which those responsible in each case set their standards and were held to account for them, and in such a way that they could even attribute things to themselves with which they had not reckoned. By using the word *computus*, Cassiodorus and Bede, Hermann the Lame and Roger Bacon, Nicholas of Cusa and Montaigne were declaring that they shared this fundamental experience.

In the last four hundred years this experience has been partly overlaid by other experiences which are reflected in the transition from *computus* to *computer*, and which signal the beginning of a new era. What is new about the modern age is not the uniqueness of events or the changeability of structures. Each generation has always justifiably felt that what happened to them and was expected of them in their lifetime was without precedent. In the modern age, it is only the acceleration of historical change, far beyond human comprehension, which has increased. Changes no longer occur gradually, between generations and regions, but within a few years and throughout the world. It is no longer just the scholars in their studies who feel the piercing wind of change, but people on the street as well. Most of these upheavals have considerably prolonged and substantially enriched human life; if its very existence is to be ensured, permanent innovations will be needed in the future too.[216]

However, in the process even the non-simultaneity of concurrent things has also become immeasurable. The rapid standardization of time-reckoning and time-measurement is matched by an equally swift fragmentation in the modern perception and usage of time. On the one hand, we search for lost time, on the other we while away the time we have saved; for some the future holds out the prospect of enjoying the most arbitrary things, while for others it offers the prospect of lacking the things they most need. Anyone who does not entrust the task of mastering this new complexity to computers finds it renewed and intensified by the old experiences of leading thinkers. Now, however, everyone faces the challenge of perceiving their time in a way that Ingeborg Bachmann described in 1953:

> Gazing into the mist, you see
> Time, borrowed until revoked,
> Come into view on the horizon.[217]

Notes

1. Norbert Elias, *Time: An Essay* (1992; first published in German 1984), pp. 193–8: calendar; p. 89: sense of time. For a more nuanced study of the concept of time, but a similar assessment of the medieval calendar, see Rudolf Wendorf, *Zeit und Kultur. Geschichte des Zeitbewusstseins in Europa* (1980), pp. 92–150.
2. Günter Dux, *Die Zeit in der Geschichte. Ihre Entwicklungslogik vom Mythos zur Weltzeit* (1989), pp. 312–48 (reference by August Nitschke). The most important prior study of the subject is Jacques Le Goff's 'The Town as an Agent of Civilisation, 1200–1500', in: Carlo M. Cipolla (ed.), *The Middle Ages* (*Fontana Economic History of Europe*, vol. 1, 1972), pp. 71–106, here p. 87. Le Goff, however, avoids positing monks in diametrical opposition to townspeople. See his introduction to *The Medieval World* (1990), pp. 17–22.
3. Thomas Nipperdey, 'Die Aktualität des Mittelalters. Über die historischen Grundlagen der Modernität', now in: Nipperdey, *Nachdenken über die deutsche Geschichte* (1989), pp. 21–30, here pp. 25ff. For different points of emphasis, but a similar view, see Ernst Schulin, 'Die historische Zeit – Dauer und Wandel', in: *Funk-Kolleg Geschichte*, ed. Werner Conze et al., 1 (1981), pp. 265–87, here pp. 269ff, 276ff.
4. Hermann Grotefend, *Taschenbuch der Zeitrechnung des deutschen Mittelalters und der Neuzeit* (1971), pp. 1–24, for a summary of the dates. For a study of their context, see Josef J. Duggan, 'The Experience of Time as a Fundamental Element of the Stock of Knowledge in Medieval Society', in: Hans Ulrich Gumbrecht et al. (eds), *La Littérature historiographique des origines à 1500*, 1 (*Grundriss der romanischen Literaturen des Mittelalters*, 11/1, 1986), pp. 127–34. For the most judicious study of the history of science, see Eugen Meyer et al., 'Chronologie', in: *Die Religion in Geschichte und Gegenwart*, 1 (1957), cols 1806–18; for a sketchy analysis, see Alfred Cordoliani, 'Comput, chronologie, calendriers', in: *L'Histoire et ses méthodes*, ed. Charles

Samaran (1961), pp. 37–51.

5. David S. Landes, *Revolution in Time. Clocks and the Making of the Modern World* (1983), pp. 6ff, 58–66, 92 (reference by Thomas Nipperdey); Wilhelm Flitner, *Die Geschichte der abendländischen Lebensformen* (1967), pp. 111ff, 197ff, 324–7.

6. As to the modes of writing, a Dutch dictionary written *c.*1480 noted that the original and linguistically correct word was *computus*, but it had to be modified to *compotus* to make it easier on the ear: Johann W. Fuchs et al. (eds), *Lexicon latinitatis nederlandicae medii aevi*, 2 (1981), col. 755. In the ninth and even in the thirteenth century, the stress on *compotus* was placed on the first syllable, following the classical convention. This is shown by the beginning of the Carolingian hexameter *compotus hic* (see below, n. 76), and the old French *contes* (below, n. 145). The Middle English word *compute* (below, n. 171) shows that the second syllable was drawn out and stressed by the fifteenth century at the latest.

7. Herbert Grundmann, 'Naturwissenschaft und Medizin in mittelalterlichen Schulen und Universitäten', now in: Grundmann, *Augewählte Aufsätze*, 3 (*Schriften der Monumenta Germaniae Historica*) [henceforth abbreviated as *MGH*], 25/3, 1978), pp. 343–67, here p. 353. The following studies are impaired by ignorance of the computus: Karl Brunner, 'Die Zeit des Menschen. Überlegungen zur Geschichte des Zeitbegriffs', in: *Das Phänomen Zeit*, ed. Manfred Horvat (1984), pp. 19–25; Hans-Werner Goetz, *Leben im Mittelalter vom 7. bis zum 13. Jahrhundert* (1986), pp. 24ff, 105ff.

8. Herman H. Goldstine, *The Computer from Pascal to von Neumann* (1972), pp. 123, 150: the origins and nomenclature of early computer models; Josef Weizenbaum, *Computer Power and Human Reason: From Judgement to Calculation* (1984), pp. 139–65: language and computers. Both overlook the etymology of the word *computer*, as do such collected volumes as Allen Kent et al. (eds), *Encyclopaedia of Computer Science and Technology*, 18 vols (1975–87), and the only specialist study to combine the superior knowledge of the computer with a thorough understanding of the computus: Heinz Zemanek, *Kalender und Chronologie. Bekanntes und Unbekanntes aus der Kalenderwissenschaft* (1987), pp. 35–60.

9. The following account differentiates the questions I raised in: 'Das mittelalterliche Zahlenkampfspiel', *Supplemente zu den Sitzungsberichten der Heidelberger Akademie der Wissenschaften*, Phil.-hist. Kl., 5 (1986), and prepares the way for my edition of the writings of Hermann the Lame on time-reckoning and time-measurement.

10. Edmund R. Leach, *Rethinking Anthropology* (1961), pp. 124–136. here p. 125. For a basic study of the conception of time in archaic societies, see Martin P. Nilsson, *Primitive Time-Reckoning. A Study in the Origins and First Development of the Art of Counting Time among the Primitive and Early Culture People* (1920), pp. 11–225; Dux (as n. 2), pp. 103–257, provides abundant evidence.

11. *Herodoti Historiae*, ed. Karl Hude, 2 vols (1927) [English translation: Herodotus, *The Histories*, 1964], unpaginated, here VIII, 51, 1: Calliades; II, 109, 3: sundial; II, 4, 1: division of the year; II, 82, 1: month and day. See Hermann Strasburger, 'Herodots Zeitrechnung', in: *Herodot. Eine Auswahl aus der neueren Forschung*, ed. Walter Marg, *Wege der Forschung*,

26 (1982), pp. 688–736, here p. 693: old wives' tales; Christian Meier, 'Die Entstehung der Historie', in: Reinhart Koselleck and Wolf-Dieter Stempel (eds), *Geschichte – Ereignis und Erzählung, Poetik und Hermeneutik*, 5 (1973), pp. 251–305, here p. 289: from zero to zero; Dux (as n. 2), pp. 273–85.

12. Plato, *Timaeus*, c. 3, in: *Opera*, ed. James Burnet, 4 (1902) [English translation: *Timaeus and Critias*, trs. H. D. P. Lee, 1965], 21e–24d: Solon; c. 10, 37c–e: the creation of time; c. 11, 39b–e: planetary orbits and temporal numbers; c. 14, 42d: instruments of time; c. 16, 47a–b: symbols and philosophy. See Hans-Georg Gadamer, 'Idee und Wirklichkeit in Platons Timaios', *Sitzungsberichte der Heidelberger Akademie der Wissenschaften*, Phil.-hist. Kl. 1974/2 (1974), pp. 14–16; Gernot Böhme, *Zeit und Zahl. Studien zur Zeittheorie bei Platon, Aristoteles, Leibniz und Kant* (1974), pp. 68–158, also examines the consequences.

13. Plato, *Timaeus*, c. 24, 59c–d: pleasure; *Politeia*, VII, 10, 527d: knowledge. See Bartel L. van der Waerden, *Die Astronomie der Griechen: Eine Einführung* (1988), pp. 34–9, 44–62. Dux (as n. 2) omits Plato.

14. Aristotle, 'Peri hermeneias', c. 9, in: *Opera*, ed. Immanuel Bekker and Olof Gigon, 1 (1960), 18a–19b. For a sceptical approach, see Harald Weinrich, *Tempus. Besprochene und erzählte Welt* (1985), pp. 55ff, 288–93. However, the notion of time was undeniably based on the tenses.

15. Aristotle, *Politics*, III, 15, vol. 2 (1960), 1286b; IV, 13, 1297b: constitution; *Metaphysics*, V, 11, 1018b: Trojan War; *Problems of Physics*, XVII, 3, 916a: people of Troy; *Poetics*, c. 9, 1451b: Herodotus; c. 23, 1459a: Homer. See Christian Meier, *Entstehung des Begriffs 'Demokratie'. Vier Prolegomena zu einer historischen Theorie* (1981), pp. 52–67; Dux (as n. 2), pp. 230ff.

16. Aristotle, *Physics*, IV, 11, vol. 1, 219b. See Wolfgang Wieland, *Die aristotelische Physik. Untersuchungen über die Grundlegung der Naturwissenschaft und die sprachlichen Bedingungen der Prinzipienforschung bei Aristoteles* (1962), pp. 316–29; Peter Janich, *Die Protophysik der Zeit. Konstruktive Begründung und Geschichte der Zeitmessung* (1980), pp. 246–59; Paul F. Conen, *Die Zeittheorie des Aristoteles* (1964), pp. 30–61, is too limited in focus.

17. Aristotle, *Categories*, c. 6, vol. 1, 5a: time and number; *Nichomachean Ethics*, II, 1, vol. 2, 1103a: master architect; I, 7, 1098a: inventor. Landes (as n. 5) overlooks the Aristotelian origins and ancient beginnings of temporal quantification.

18. Aristotle, *Metaphysics*, X, 1, vol. 2, 1053e: movement of the heavens; *Problems of Physics*, XV, 5–10, 911a–912b: shadow. For the best study of Aristotle's influence, see Alexandre Koyré, *Galilei. Die Anfänge der neuzeitlichen Wissenschaft* (1988), pp. 13–28.

19. Otto Neugebauer, *A History of Ancient Mathematical Astronomy*, 3 vols (1975), here vol. 3, pp. 1061–76 on the astronomical principles of historical chronology; vol. 1, pp. 353–66 on the Babylonian calendar; vol. 2, pp. 559–68 on the Egyptian solar year and lunar cycle. For a study of the Jewish lunar calendar, see Eduard Mahler, *Handbuch der jüdischen Chronologie* (1916), pp. 17–59; Ludwig Basnizki, *Der jüdische Kalender. Entstehung und Aufbau* (1986), pp. 9–32.

20. Waerden (as n. 13), pp. 76–92 provides a summary of the Greeks'

astronomical calendar. For a study of their clocks, see Hermann Diels, *Antike Technik* (1924), pp. 155–228; for a complementary study of water-clocks, see Aage G. Drachmann, *The Mechanical Technology of Greek and Roman Antiquity* (1963) pp. 192ff; on sundials, see Edmund Buchner, 'Antike Reiseuhren', *Chiron*, 1 (1971), pp. 457–82. On the prehistory of the astrolabe, see Waerden (as n. 13), pp. 101–4; for a different approach, see Borst, 'Astrolab und Klosterreform an der Jahrtausendwende', *Sitzungsberichte der Heidelberger Akademie der Wissenschaften*, Phil.-hist. Kl. 1989/1 (1989), pp. 13–19.

21. Both the following note a conflict between cyclical and linear time: Karl Löwith, *Weltgeschichte und Heilsgeschehen. Die theologischen Voraussetzungen der Geschichtsphilosophie* (1967), p. 26; Theodor Schieder, *Geschichte als Wissenschaft. Eine Einführung* (1968), pp. 81ff. This view is justifiably opposed by Arnaldo Momigliano, 'Time in Ancient Historiography', *History and Theory*, Suppl. VI (1966), pp. 1–23, and Friedrich Vittinghoff, 'Spätantike und Frühchristentum. Christliche und nichtchristliche Anschauungsmodelle', in: *Mensch und Weltgeschichte. Zur Geschichte der Universalgeschichtsschreibung*, ed. Alexander Randa (1969), pp. 17–40.

22. Eviatar Zerubavel, *The Seven Day Circle. The History and Meaning of the Week* (1985), pp. 5–26 on the origins; for a discussion of its influence on the Middle Ages, see Georg Schreiber, *Die Wochentage im Erlebnis der Ostkirche und des christlichen Abendlandes* (1959), pp. 20–43.

23. Elias J. Bickermann, *Chronology of the Ancient World* (1980), pp. 43–51, examines Roman chronology in general. On ancient Roman chronology, see Agnes K. Michels, *The Calendar of the Roman Republic* (1967). On Caesar's reform, see Wilhelm Kubitschek, *Grundriss der antiken Zeitrechnung* (1928), pp. 99–109; Christian Meier, *Caesar* (1982; English translation forthcoming), pp. 528ff. Dux (as n. 2) overlooks this first attempt at establishing a universal time. For the best study of Vitruvius's clocks, see the commentary to his ninth book: *Vitruve De l'architecture livre IX*, ed. Jean Soubiran (1960), pp. 214–308.

24. Edmund Buchner, *Die Sonnenuhr des Augustus* (1982), pp. 7–80 on the obelisk on the Campus Martius; for more on the Vatican obelisk, see Geza Alföldy, 'Der Obelisk auf dem Petersplatz im Rom', *Sitzungsberichte der Heidelberger Akademie der Wissenschaften*, Phil.-hist. Kl. 1990/2 (1990), pp. 55–67. For its later reinterpretation, see n. 133 below. For the Augustan conception of time, see Hubert Canick, 'Die Rechtfertigung Gottes durch den "Fortschritt der Zeiten"', in: *Die Zeit. Dauer und Augenblick*, ed. Armin Mohler et al. (1989), pp. 257–88, here pp. 265–81.

25. For the most vivid surveys of early Christian times, see Charles W. Jones, 'Development of the Latin Ecclesiastical Calendar', in *Bedae Opera de temporibus*, ed. Charles W. Jones (The Medieval Academy of America Publication 41, 1943), pp. 1–122, here pp. 6–68, especially on the decrees of 325, pp. 17–25; August Strobel, *Ursprung und Geschichte des frühchristlichen Osterkalenders* (1977), pp. 122–394, on Nicaea, pp. 389–92. Wesley Stevens is preparing a two-volume catalogue of all literature on time-reckoning from 200 to 1582.

26. Plautus, *Miles gloriosus*, 204, in: *Comoediae*, ed. Friedrich Leo, 2 (1958), p.

15: 'dextera digitis rationem computat.' On the classical semantic field of *computare, computatio* and *computus*, see Bertold Maurenbrecher, in: *Thesaurus linguae Latinae*, 3 (1912), cols 2175–86. On mathematical terminology, see Johannes Tropfke, *Geschichte der Elementarmathematik*, 1 (1980), pp. 34, 122, 168. On arithmetical practice, see Anita Rieche, 'Computatio Romana. Fingerzählen auf provinzialrömischen Reliefs', *Bonner Jahrbücher*, 186 (1986), pp. 165–92 (reference by Ute Schillinger).

27. Irenaeus of Lyon, *Adversus haereses*, I, 15, 2, ed. Adelin Rousseau and Louis Doutreleau, 1/2 (*Sources chrétiennes*, 264, 1979), pp. 236–8 added together the numerical values of the Greek letters in the name Jesus and called the total *arithmos*; the Latin translation gives it as *computus*. Whether it was done in the third century or not until the fifth century is a matter of dispute. Pseudo-Cyprian of Carthage, *De pascha computus*, *Corpus scriptorum ecclesiasticorum Latinorum*, 3 (1871), pp. 248–71, wrote in 243, but in the context used only *computare*, never *computus*, c. 4, p. 251, and elsewhere. A Reims manuscript of the ninth century was the first to formulate the book title of the edition.

28. Julius Firmicus Maternus, *Mathesis*, I, 4, 5, ed. Wilhelm Kroll and Frank Skutsch (1968), p. 12. For an accurate account of the reform, see Charles Ducange and Leopold Favre, *Glossarium mediae et infimae latinitatis*, 2 (1883), p. 473.

29. Jerome, *Chronicon*, a. Abr. 985, *Die griechischen christlichen Schriftsteller*, 24 (1913), fo. 70, also a. 361, fo. 36. See Anna-Dorothee von den Brincken, *Studien zur lateinischen Weltchronistik bis in das Zeitalter Ottos von Freising* (1957), pp. 60–7. For a discussion of the Jewish world era with the epoch of 7 October 3761 BC which was conceived no later than the fourth century, but not accepted until the twelfth century, see Mahler (as n. 19), pp. 153–9, 455–79 (reference by Alexander Patschovsky).

30. Augustine, *Confessiones*, IV, 16, 28ff, *Corpus Christianorum Series Latina*, 27 (1981), p. 54: categories; XI, 18, 23, p. 205: images. Janich (as n. 16), pp. 259–71, is too limited; Dux (as n. 2), pp. 322–7, is similarly too narrow, though in different respects. I go along with Ernst A. Schmidt, 'Zeit und Geschichte bei Augustin', *Sitzungsberichte der Heidelberger Akademie der Wissenschaften*, Phil.-hist. Kl. 1985/3 (1985), pp. 17–32.

31. Augustine, *Confessions*, XI, 20, 26, p. 206ff: three times; XI, 28, 38, p. 214: singer; XI, 23, 29, pp. 208ff: heavenly bodies; XI, 24, 31, p. 210: movement.

32. 'Contra Felicem' I, 10, *Corpus scriptorum ecclesiasticorum Latinorum*, 25/2 (1892), p. 812.

33. *De civitate Dei* XI, 30, *Corpus Christianorum Series Latina*, 48 (1955), pp. 350ff: the number six; the quotation is from Wisdom 11:21; XVIII, 52ff, pp. 650–2: persecutions of the Christians; XXII, 30, pp. 865ff: the number seven. See Reinhart Koselleck, *Future Past. On the Semantics of Historial Time* (1985); Schmidt (as n. 30), pp. 96–109.

34. Augustine, *Epistula* 199, 34, *Corpus scriptorum ecclesiasticorum Latinorum*, 57 (1911), pp. 273ff.

35. Boethius, *De institutione arithmetica*, I, 2, ed. Gottfried Friedlein (1867), p. 12: exemplar; 1, 1, pp. 8ff: arithmetic and astronomy. For this arithmetic and its consequences, see Detlef Illmer, 'Arithmetik in der

gelehrten Arbeitsweise des frühen Mittelalters. Eine Studie zum Grundsatz "Nisi enim nomen scieris, cognito rerum perit"', in: *Institutionen, Kultur and Gesellschaft im Mittelalter. Festschrift für Josef Fleckenstein* (1984), pp. 35–58; Menso Folkerts, 'Die Bedeutung des lateinischen Mittelalters für die Entwicklung der Mathematik. Forschungsstand und Probleme', now in: *Wissenschaftsgeschichte heute*, ed. Christian Hünemörder (1987), pp. 87–114.

36. Boethius, *De institutione musica*, I, 2, ed. Gottfried Friedlein (1867), pp. 187ff: harmony of the spheres and seasons; II, 8, p. 234; II, 29, p. 263: *computare*. For this music theory and its consequences, see Michael Bernhard, 'Überlieferung und Fortleben der antiken lateinischen Musiktheorie im Mittelalter', in: *Geschichte der Musiktheorie*, 3, ed. Frieder Zaminer (1990), pp. 7–35, here pp. 24–31.

37. *Epistula Theophili*, c. 2, ed. Bruno Krusch, *Studien zur christlich-mittelalterlichen Chronologie. Der 84–jährige Ostercyclus und seine Quellen* (1880), p. 221. The *Epistula Proterii*, cc. 6–7 (translated by Dionysius himself), ibid., p. 275, already made a distinction between *dominicum pascha* and *paschalis conpotus*.

38. Dionysius Exiguus, *Libellus de cyclo magno paschae*, ed. Bruno Krusch, *Studien zur christlich-mittelalterlichen Chronologie. Die Entstehung unserer Zeitrechnung*, *Abhandlungen der Preussischen Akademie der Wissenschaften*, Phil.-hist. Kl. 1937/8 (1938), pp. 63ff. See Jones (as n. 25), pp. 68–75; for a different perspective on the subject, see Walter E. van Wijk, *Origine et développement de la computistique médiévale* (1954), pp. 15ff.

39. *Benedicti Regula*, cc. 16–18, *Corpus scriptorum ecclesiasticorum Latinorum*, 75 (1977) [English translation: *The Rule of Saint Benedict*, trs. D. Parry, 1984], pp. 70–81: canonical hours and order of psalms; c. 8, p. 58: time of rising; c. 41, pp. 112–14: meal times; c. 48, pp. 125–8: hours of work and rest, here pp. 126ff. 'From Easter . . .', p. 125: idleness; 'Prologue', p. 7: time to make peace. See Gustav Bilfinger, *Die mittelalterlichen Horen und die modernen Stunden. Ein Beitrag zur Kulturgeschichte* (1892), pp. 1–7, 109–25; Stephen M. McCluskey, 'Gregory of Tours, Monastic Timekeeping, and Early Christian Attitudes to Astronomy', *Isis*, 81 (1990), pp. 9–22, here pp. 9ff, 19ff. For a challenge to the usual misinterpretation of the Benedictine work ethic, see Friedrich Prinz, *Askese und Kultur. Vor- und frühbenediktinisches Mönchtum an der Wiege Europas* (1980), pp. 68–74; Dux (as n. 2), pp. 320–2.

40. Cassiodorus Senator, *Institutiones*, II, 4, 7, ed. Roger A. B. Mynors (1961), p. 141: arithmetic; II, 7, 3–4, p. 156: astronomy. See Heinz Löwe, 'Cassiodor', now in: Löwe, *Von Cassiodor zu Dante. Ausgewählte Aufsätze zur Geschichtsschreibung und politischen Ideenwelt des Mittelalters* (1973), pp. 11–32, here pp. 22–8. Alexander Murray, *Reason and Society in the Middle Ages* (1978), pp. 145, 154, undervalues Cassiodorus's achievements in arithmetic and computus.

41. Cassiodorus Senator, *Institutiones*, I, 30, 5, pp. 77ff: on the monks; to Boethius: *Variae*, I, 45, *MGH Auctores antiquissimi*, 12 (1894), pp. 39–41. On the various meanings of *horologium*, see Landes (as n. 5), pp. 53, 68. Like McCluskey (as n. 39), he overlooks Cassiodorus's opinions, and in general neglects the early medieval change from time-measurement to

time-reckoning.

42. *Computus paschalis*, ed. Paul Lehmann, *Cassidorstudien*, now in: Paul Lehmann, *Erforschung des Mittelalters. Ausgewählte Abhandlungen und Aufsätze*, 2 (1959), pp. 38–108, here pp. 52–5. A list of the earliest computistical writings is given in: Eloi Dekkers and Emile Gaar, *Clavis patrum Latinorum* (1961), pp. 507–18.

43. Gregory I, *Homiliae in Hiezechielem II*, I, 5, 12, *Corpus Christianorum Series Latina*, 142 (1971), p. 285. See Heinz Meyer, *Die Zahlenallegorese im Mittelalter. Methode und Gebrauch* (1975), pp. 32–4. An overgeneralized account is given by Franz Carl Endres and Annemarie Schimmel, *Das Mysterium der Zahl. Zahlensymbolik im Kulturvergleich* (1984), pp. 33–5.

44. Gregory I, *Homiliae in Evangelia*, I, 19, 1–2, ed. Jacques-Paul Migne, *Patrologia latina*, 76 (1851), col. 1154f: on Matthew 20:1–16. See Roderich Schmidt, 'Aetates Mundi. Die Weltalter als Gliederungsprinzip der Geschichte', *Zeitschrift für Kirchengeschichte*, 67 (1956), pp. 288–317, here pp. 302ff.

45. Gregory of Tours, *Libri historiarum IV*, 17, *MGH Scriptores rerum Merovingicarum*, 1/1 (1951), p. 215, and X, 23, pp. 514ff: Easter doubts and Easter miracle; IV, 46, p. 181: slave; IV, 51, pp. 189ff and X, 31, pp. 536ff: enumeration of years; I, 'Praefatio', p.5: *conpotare*. For the Easter miracle, see Charles W. Jones, 'A Legend of St. Pachomius', *Speculum*, 18 (1943), pp. 198–210, here p. 207. For Gregory's belief in the miracle, see Aaron Gurevich, *Mittelalterliche Volkskultur. Probleme zur Forschung* (1986), pp. 32–6, 39–42. On the decline of the art of arithmetic, see Murray (as n. 40), p. 144.

46. Gregory of Tours, *De cursu stellarum ratio*, *MGH Scriptores rerum Merovingicarum*, 1/2 (1969), pp. 407–22, here c. 16, p. 413: the quotation. See Werner Bergmann and Wolfhard Schlosser, 'Gregor von Tours und der "rote Sirius". Untersuchungen zu den astronomischen Angaben in "De cursu stellarum"', *Francia*, 15 (1987), pp. 43–74; McCluskey (as n. 39), pp. 10–19, deviates from this in some details.

47. Isidore of Seville, *Etymologiae*, III, 4, 3–4, ed. Wallace M. Lindsay, 1 (1911, unpaginated): *conputus*. The preceding sentences, from III, 4, 1–2, summarize Augustine's ideas (as above, n. 33); XX, 13, 5, vol. 2 (1911): *horologia*. See Arno Borst, 'Das Bild der Geschichte in der Enzyklopädie Isidors von Sevilla', *Deutsches Archiv für Erforschung des Mittelalters* (henceforth abbreviated as *Deutsches Archiv*), 22 (1966), pp. 1–62, here pp. 13–15. Landes (as n. 5), p. 64, wrongly dismisses Isidore's notions of time as rudimentary.

48. Isidore, *Etymologiae*, III, 5, 10, vol. 1: 'to add'; XVI, 25, 19, vol. 2: 'multiply'; I, 3, 1, vol. 1 and X, 43: *calculator*; V, 29, 1–36, 3: from the moment to the Great Year; V, 38, 3–5 and V, 39, 1: sequences of dates and the number six; VI, 17, 15–18: computation of Easter; V, 35, 1: *tempora*. Later evidence of the derivation of *temperamentum* is cited by Jean Leclercq, 'Experience and Interpretation of Time in the Early Middle Ages', *Studies in Medieval Culture*, 5 (1975), pp. 9–19, here p. 16.

49. *De ratione computandi*, c. 3, ed. Maura Walsh and Dáibhí Ó Cróinin, *Dummian's Letter de controversia paschali* (1988), pp. 117ff. See Dáibhí Ó Cróinin, 'A Seventh-Century Irish Computus from the Circle of Cummianus', *Proceedings of the Royal Irish Academy*, 82/C/11 (1982), pp.

405–30, here p. 411 (reference by Michael Richter). This discovery is left out of account by Knut Schäferdiek, 'Der irische Osterzyklus des sechsten und siebten Jahrhunderts', *Deutsches Archiv*, 39 (1983), pp. 357–83, where later Irish computi are discussed.

50. Pseudo-Fredegar, *Chronicon*, I, 24, *MGH Scriptores rerum Merovingicarum*, 2 (1988), p. 34, and III, 73, pp. 112ff: *supputatio*; II, 7, p. 47: Samson. For a recent discussion of the work, see Andreas Kusternig, 'Einleitung', *Freiherr vom Stein – Gedächtnisausgabe*, 4a (1982), pp. 1–33.

51. *MGH Scriptores rerum Merovingicarum*, 7 (1920), p. 499.

52. 'Der merovingische Computus Paschalis vom Jahre 727 n. Chr.', ed. Krusch (as n. 38), pp. 53–7. See Alfred Cordoliani, 'Les plus anciens manuscrits de comput ecclésiastique de la bibliothèque de Berne', *Zeitschrift für schweizerische Kirchengeschichte*, 51 (1957), pp. 101–12.

53. John Henning, 'Kalender und Martyrologium als Literaturformen', now in: Hennig, *Literatur und Existenz. Ausgewählte Aufsätze* (1980), pp. 37–80 on the liturgical concept of time. For a discussion of the holy day, see Hans Martin Schaller, 'Der heilige Tag als Termin mittelalterlicher Staatsakte', *Deutsches Archiv*, 30 (1974), pp. 1–24, here p. 23. For a study of the medieval concept of time, see Aaron J. Gurevich, *Das Weltbild des mittelalterlichen Menschen* (first published in Russian as *Kategorii Srednevekovoi kul'tury*) (1980), pp. 98–122.

54. Bede, *Epistola ad Wicthedum*, c. 6, in: *Opera didascalica*, ed. Charles W. Jones, *Corpus Christianorum Series latina*, 123, A–C (1975–80), here vol. C, p. 637; c. 12, p. 642: equinox. *De natura rerum*, c. 47–8, vol. A, pp. 229–32: geographical latitude. *De temporum ratione*, c. 38, vol. B, pp. 400ff: leap-day. See Ernst Zinner, *Alte Sonnenuhren an europäischen Gebäuden* (1964), pp. 2ff; Wesley M. Stevens, *Bede's Scientific Achievement* (Jarrow Lecture, 1985), pp. 5–10, 24ff, 43ff.

55. Bede, *De temporum ratione*, c. 1, vol. B, p. 268. For a fundamental study of the entire work, see Charles W. Jones, 'The Computistical Works of Bede', in: Charles W. Jones (as n. 25), pp. 123–72, supplemented by his 'Foreword', vol. A, pp. xii–xvi; this is extended by Murray (as n. 40), pp. 146–51. On the passage, see Alfred Cordoliani, 'A propos du chapitre premier du "De temporum ratione" de Bède', *Le moyen âge*, 54 (1948), pp. 209–23.

56. Bede, *De temporum ratione*, c. 38, p. 400: *calculator* and *computator*; c. 11, p. 317: *computare* and *calculare* as synonyms. *Historia ecclesiastica gentis Anglorum*, III, 25, ed. Bertram Colgrave and Roger A. B. Mynors, *Bede's Ecclesiastical History of the English People* (1969), p. 306: *catholicus calculator*; IV, 2, p. 332: *arithmetica ecclesiastica*. See Murray (as n. 40), p. 148; Landes (as n. 5), p. 64. The most concise study of the semantic field of *computare*, *computatio*, and *computus* centred on Bede is provided by Irmengard Dauser, in: *Mittelalterliches Wörterbuch*, 2 (1985), cols 1128–34.

57. Bede, *De temporum ratione*, c. 19, vol. 13, pp. 343–6 and c. 23, pp. 353–5: tables with letters; *De temporibus*, c. 12, vol. C, p. 595: against *calculandi facilitas*; *De temporum ratione*, c. 38, vol. B, p. 399: against *facilitas computandi*; cc. 41–3, pp. 405–18: 'lunar jump'. Alistair C. Crombie, *From Augustine to Galileo* (1961), pp. 25–8 offers a mistakenly modern interpretation of Bede's methods as practical empiricism. Bede was still less thinking of a reform of the Julian Calendar, though Zemanek (as n.

8), p. 29 implies that he was.

58. Bede, *De temporum ratione*, 'Praefatio', p. 263: God; c. 2, pp. 274ff: the three types of chronology.

59. Ibid., c. 3, pp. 276–8: astrologers, the hour and the *horologium*; c. 5, pp. 283ff: twenty-four hours. Bede is omitted from the history of equal hours surveyed by Igor A. Jenzen, *Uhrzeiten. Die Geschichte der Uhr und ihres Gebrauches* (1989), pp. 31–6.

60. Bede, *De temporum ratione*, c. 6, pp. 290–5: day and month of the Creation; c. 66, pp. 495ff: its year. See Anna-Dorothee von den Brincken, 'Weltären', *Archiv für Kulturgeschichte*, 39 (1957), pp. 133–49, here pp. 146ff.

61. Bede, *De temporum ratione*, c. 66, pp. 463–535, according to Theodor Mommsen (ed.), *MGH Auctores antiquissimi*, 13 (1898), pp. 247–321. See Arno Borst, 'Universal Histories in the Middle Ages?' now in: Borst, *Medieval Worlds* (1991; first published in German, 1988), pp. 63–71.

62. Bede, *De temporum ratione*, c. 15, p. 331. See Jacob and Wilhelm Grimm, *Deutsches Wörterbuch*, 13 (1889), cols 1371ff. For more on Bede's comparison, see Arno Borst, *Lebensformen im Mittelalter* (1988), pp. 35–49.

63. Bede, *Historia* (as n. 56), I, 4, p. 24; V, 24, pp. 560–6, and other references. In I, 2, p. 20, dating was also in years BC. See Anna-Dorothee von den Brincken, 'Beobachtungen zum Aufkommen der retrospektiven Inkarnationsära', *Archiv für Diplomatik*, 25 (1979), pp. 1–20, here p. 16. Dux (as n. 2) overlooks the fact that the establishment of this era has determined the dating of our universal time to the present day.

64. Bede, *Historia*, V, 24, p. 570. For a fundamental account, see Henri Quentin, *Les Martyrologes historiques du moyen âge* (1908), pp. 17–119; see also John McCullogh, 'Historical Martyrologies in the Benedictine Cultural Tradition', in: *Benedictine Culture 750–1050*, ed. Willem Lourdeaux and Daniel Verhelst (1983), pp. 114–31.

65. Bede, *Martyrologium*, ed. Quentin (as n. 64), p. 109; *De temporum ratione*, c. 66, vol. B, p. 535.

66. In addition to Bede's computation, Murray (as n. 40) lends validity only to his historiography, pp. 149ff. Franz-Josef Schmale, *Funktion und Formen mittelalterlicher Geschichtsschreibung* (1980), pp. 28–37, includes an account of computation but not of the martyrologies; likewise Karl Heinrich Krüger, Die Universalchroniken. *Typologie des sources du moyen âge occidental*, 16 (1976), pp. 13–21; supplement (1985), pp. 1ff. The connection is completely broken in Bernard Guenée's book, *Histoire et culture historique dans l'Occident médiéval* (1980), pp. 52–4.

67. Boniface, *Epistola 76, MGH Epistolae selectae*, 1 (1955), p. 159: *clocca*. Walahfrid Strabo, *Vita sancti Galli*, II, 10, *MGH Scriptores rerum Merovingicarum*, 4 (1910), p. 320: hand-bell; II, 4, p. 315: tower-bell, here *campanum* (neuter). Walahfrid Strabo, *De exordiis et incrementis quarundam in observationibus ecclesiasticis rerum*, c. 5, *MGH Capitularia regum Francorum*, 2 (1897), pp. 478ff.: origin of the word *campana* (feminine). On the word *Glocke*, see Grimm (as n. 62), vol. 8 (1958), cols 142ff. On the object, see Landes (as n. 5), pp. 68ff; Kurt Kramer, 'Glocke', in: *Lexikon des Mittelalters*, 4 (1989), cols 1497–1500.

68. *Admonitio generalis*, c. 72, *MGH Capitularia regum Francorum*, 1 (1883), p.

60, no. 22; briefly repeated on p. 121, no. 43; p. 235, no. 117; p. 237, no. 119. Similarly, on Charlemagne's lifetime, see Haito of Basle, *Capitula*, c. 6, *MGH Capitula episcoporum*, 1 (1984), p. 211 and Waltcaud of Liège, c. 11, ibid., p. 47. Thereafter, 827, Abbot Ansegis of Fontenelle, *Collectio capitularium*, *MGH Capitularia regum Francorum*, 1, pp. 403, 446. A register of Carolingian writings on computation is given in Alfred Cordoliani, 'Les traités de comput du haut moyen âge 526–1003', *Archivum latinitatis medii aevi*, 17 (1942), pp. 51–72, which has been superseded in several respects but not replaced.

69. Einhard, *Vita Karoli magni*, c. 25, *MGH Scriptores rerum Germanicarum*, 25 (1911), p. 30: computus. *Annales regni Francorum*, a. 807, *MGH Scriptores rerum Germanicarum*, 6 (1895), pp. 123ff: water-clock. See Percy Ernst Schramm, 'Karl der Grosse. Denkart und Grundauffassungen', now in: Schramm, *Kaiser, Könige und Päpste. Gesammelte Aufsätze zur Geschichte des Mittelalters*, 1 (1968), pp. 302–41, here pp. 311–27; Murray (as n. 40), p. 151. An evaluation of the water-clock is provided by Landes (as n. 5), p. 24.

70. Alcuin, *Epistola 171*, *MGH Epistolae*, 4 (1895), pp. 281–3: *computus* and *calculatio*; *Epistola 145*, pp. 231–5: *calculatores* and *mathematici*; cf. *Epistola 126*, pp. 185–7. A cautious study of Alcuin's mathematical interests is carried out by Menso Folkerts in: 'Die älteste mathematische Aufgabensammlung in lateinischer Sprache: Die Alkuin zugeschriebenen Propositiones ad acuendos iuvenes', *Österreichische Akademie der Wissenschaften*, Math.-nat. Kl. Denkschriften, 116/6 (1978), pp. 30ff.

71. Einhard, *Vita Karoli magni*, c. 29 (as n. 69), p. 33. See Dieter Geuenich, 'Die volkssprachliche Überlieferung der Karolingerzeit aus der Sicht des Historikers', *Deutsches Archiv*, 39 (1983), pp. 104–30, here pp. 124–7. For Bede's example, see above n. 62.

72. *MGH Epistolae*, 4, pp. 565–7. The purportedly older instance of *computista* in Jan F. Niermeyer, *Mediae latinitatis lexicon minus* (1976), p. 233, is eleventh-century (see below, n. 112). The scholarly endeavours include a Carolingian collection based on the Irish model (see above, n. 49) in equating *numerus* with *compotus*; the Prologue was edited by Alfred Cordoliani, 'Une encyclopédie carolingienne de comput. Les "Sententiae in laude compoti"', *Bibliothèque de l'École des Chartes*, 104 (1943), pp. 237–43, here p. 242. The most detailed studies of the encyclopaedia of 809 are carried out by Wilhelm Neuss, 'Ein Meisterwerk der karolingischen Buchkunst aus der Abtei Prüm', in: *Spanische Forschungen der Görres-Gesellschaft*, I/8 (1940), pp. 37–64, and Vernon H. King, *An Investigation of Some Astronomical Excerpts from Pliny's Natural History found in Manuscripts of the Earlier Middle Ages* (1969), pp. 28–53.

73. Hrabanus Maurus, *De computo*, c. 69, *Corpus Christianorum Continuatio mediaevalis*, 44 (1979), p. 284: year of the Lord; c. 65, p. 282: year of Emperor Louis; c. 68, p. 284: 22 July; *Martyrologium*, ibid., p. 113: Rufus. *De computo*, c. 36, p. 247: Great Year; c. 51, pp. 261ff: *horoscopus* and *calculator*; c. 17, p. 221: sundial; c. 11, pp. 218ff: atom; c. 8, pp. 214ff: scruple. Rash conclusions on the computus are reached by Murray (as n. 40), p. 152; a detailed study is provided by Wesley M. Stevens,

'Compotistica et Astronomica in the Fulda School', in: *Saints, Scholars and Heroes. Studies in Medieval Culture*, ed. Margot H. King and Wesley M. Stevens (1979), pp. 27–63; Maria Rissel, 'Hrabans Liber de computo als Quelle der Fuldaer Unterrichtspraxis in den Artes Arithmetik und Astronomie', in: *Hrabanus Maurus und seine Schule*, ed. Winfried Böhne (1980), pp. 138–55. On the martyrology, see John McCulloh, 'Hrabanus Maurus' Martyrology. The Method of Composition', *Sacris erudiri*, 23 (1978/9), pp. 417–61.

74. Walahfrid Strabo, *Visio Wettini*, 183–8, *MGH Poetae Latini medii aevi*, 2 (1884), p. 310: an incorrect conversion (Saturday fell on 29, not 30, October 824). Computational rules in: *Carmen LXXXIX*, ibid., pp. 422ff, unnecessarily shifted to the works of dubious authenticity. The draft was for no.1: Hrabanus, *De computo*, c. 53 (as n. 73), p. 265; for no. 2: c. 34, p. 243; for no. 3: c. 59, p. 272; for no. 4: c. 83, p. 303. See also Wesley M. Stevens, 'Walahfrid Strabo. A Student at Fulda', in: *Historical Papers 1971 of the Canadian Historical Association* (1972), pp. 13–20. On Walahfrid's computational notes, see Bernhard Bischoff, 'Eine Sammelhandschrift Walahfrid Strabos (Cod. Sangall. 878)', in: Bischoff, *Mittelalterliche Studien. Ausgewählte Aufsätze zur Schriftkunde und Literaturgeschichte*, 2 (1967), pp. 34–51, here pp. 38–41.

75. Wandalbert of Prüm, *Epistola*, *MGH Poetae Latini medii aevi*, 2, p. 569: declaration of intent; *De creatione mundi*, pp. 621ff: universal machine; *Martyrologium*, p. 582: date of the Creation; p. 597: Münstereifel. See John Hennig, 'Versus de mensibus', *Traditio*, 11 (1955), pp. 65–90; Ludolf Kuchenbuch, *Bäuerliche Gesellschaft und Klosterherrschaft im 9. Jahrhundert. Studien zur Sozialstruktur der Familia der Abtei Prüm* (1978), pp. 36ff, 107. On memory improvement, see Pierre Riché, 'Le Rôle de la mémoire dans l'enseignement médiéval', in: *Jeux de mémoire. Aspects de la mnémotechnie médiévale*, ed. Bruno Roy and Paul Zumthor (1985), pp. 133–48.

76. Agius of Corvey, *Versus computistici*, no. 2, *MGH Poetae Latini medii aevi*, 4/3 (1923), p. 939: the hexameters; the tables, cols 1178ff; no. 1, pp. 937ff: the celebration of numbers. For a detailed study, see Ewald Könsgen, 'Agius von Corvey', in: *Die deutsche Literatur des Mittelalters. Verfasserlexicon*, 1 (1978), cols 78–82. For the edition of the verses of 863, see Ewald Könsgen, 'Eine neue komputistische Dichtung des Agius von Corvey', *Mittellateinisches Jahrbuch*, 14 (1979), pp. 66–75.

77. *Annales Fuldenses*, a. 884. *MGH Scriptores rerum Germanicarum*, 7 (1891), p. 112. On the author, see Hagen Keller, 'Zum Sturz Karls III', *Deutsches Archiv*, 22 (1966), pp. 333–84. Literary and historical research today makes a clear distinction between narrated and enumerated time, for instance Weinrich (as n. 14), pp. 46–50, 136–9; Koselleck (as n. 33), pp. 144–57. This is not true of the early Middle Ages.

78. Ado of Vienne, *Martyrologium*, ed. Quentin (as n. 64), p. 636: All Saints' Day. This contains, on pp. 466–674, the best analysis of the work. On the chronicle, see Fritz Landsberg, *Das Bild der alten Geschichte in mittelalterlichen Weltchroniken*, Diss. phil. Basle (1934), pp. 33–6; Brincken (as n. 29), pp. 126–8.

79. *Le Martyrologe d'Usuard*, ed. Jacques Dubois (1965), pp. 332ff. See Jacques Dubois, 'Les Martyrologes du moyen âge latin', *Typologie des*

sources du moyen âge occidental, 26 (1978), pp. 45–56.

80. Hincmar of Reims, *Capitula synodica* (of 852), c. 8, ed. Jacques-Paul Migne, *Patrologia Latina*, 125 (1852), col. 775; *necessarius*: Hincmar, *Collectio de ecclesiis et capellis* (c. 858), *MGH Fontes iuris Germanici antiqui*, 14 (1990), p. 101, and Riculf of Soissons, *Statuta* (of 889), c. 5, ed. Migne, *Patrologia Latina*, 131 (1853), col. 17: *memoriter*. The requirements are individually listed in the Ordo Synodalis, *Inquisitio*, c. 7, ed. Carlo de Clercq, *La Législation religieuse franque*, 2 (1958), p. 410.

81. Helpericus, *Liber de computo*, 'Praefatio', ed. Jacques-Paul Migne, *Patrologia Latina*, 137 (1854), col. 17: ars compoti; 'Prologus', col. 19: *calculatoria ars*; both passages also in *MGH Epistolae*, 6 (1925), pp. 117, 119; c. 30, col. 40: appearance to the eye; c. 18, cols 32ff: lunar orbit. On the dating, see Patrick McGurk, 'Computus Helperici. Its Transmission in England in the Eleventh and Twelfth Centuries', *Medium Aevum*, 43 (1974), pp. 1–5.

82. Regino of Prüm, *De synodalibus causis*, 'Notitia' no. 93, ed. Friedrich W. Wasserschleben (1840), p. 26: *compotus minor*. Very similar, but with the notable omission of the *compotus maior*: *Commonitorium cuiusque episcopi*, c. 47, ed. Jacques-Paul Migne, *Patrologia Latina*, 96 (1851), col. 1380. Regino, *Cronicon*, 'Praefatio', *MGH Scriptores rerum Germanicarum*, 50 (1890), p. 1: the year of writing the draft; a. 718, pp. 37–40: synchronizing the cycles. See Heinz Löwe, 'Regino von Prüm und das historische Weltbild der Karolingerzeit', now in: Löwe (as n. 40), pp. 149–79, here pp. 171–4; Murray (as n. 40), pp. 152ff, 451.

83. Aurelianus Reomensis, *Musica disciplina*, c. 8, ed. Lawrence Gushee (*Corpus scriptorum de musica*, 21, 1975), p. 80, in my view misplaced in the notes of variants. On the draft, unknown to the editor, see n. 49 above. On the sung mnemonic verses, see Wolfgang Irtenkauf, 'Der Computus ecclesiasticus in der Einstimmigkeit des Mittelalters', *Archiv für Musikwissenschaft*, 14 (1957), pp. 1–15.

84. Notker Balbulus, *Martyrologium*, ed. Jacques-Paul Migne, *Patrologia Latina*, 131 (1853), col. 1114; *in hoc ecclesiasticarum historiarum brevario*; col. 1070: Marcus and George; cols 1132ff: Afra. For a standard analysis, see Ernst Dümmler, 'Das Martyrologium Notkers und seine Verwandten', *Forschungen zur Deutschen Geschichte*, 25 (1885), pp. 195–220, here pp. 202–8.

85. Marc Bloch, *The Feudal Society*, trs. L. A. Manyon (1989), pp. 73ff, gave a valid description in 1939 of the relationship of laypeople to time and numbers. The experts' knowledge is underestimated by George Duby, *The Age of the Cathedrals: Art and Society 980–1420*, trs. E. Levieux and B. Thompson (1981), pp. 47ff, 141ff. Both aspects are examined by Murray (as n. 40), pp. 157–67.

86. Johannes Fried, 'Endzeiterwartung um die Jahrtausendwende', *Deutsches Archiv*, 45 (1989), pp. 381–473. On the history of science, which the latter only touches upon, see Borst (as n. 20), pp. 13–30. On the effects this had upon monastic daily life, see Josef Semmler, 'Das Erbe der karolingischen Klosterreform im 10. Jahrhundert', in: *Monastische Reformen im 9. und 10. Jahrhundert*, ed. Raymund Kottje and Helmut Maurer, *Vorträge und Forschungen*, 38 (1989), pp. 29–78. On earlier forecasts by Franconian computists, see above, n. 52.

87. Abbo of Fleury, *Praefatio commentarii in cyclum Victorii*, ed. Jacques-Paul Migne, *Patrologia Latina*, 139 (1854), col. 572, which reads *calculatoris* instead of *calculatorii*. See also Gillian R. Evans and Alison M. Peden, 'Natural Science and the Liberal Arts in Abbo of Fleury's Commentary on the Calculus of Victorius of Aquitaine', *Viator*, 16 (1985), pp. 109–27, here the water-clock, pp. 119ff; a supplementary study is carried out by McCluskey (as n. 39), pp. 20ff.

88. Abbo, fragment and letter, ed. Alfred Cordoliani, 'Abbon de Fleury, Hériger de Lobbes et Gerland de Besançon sur l'ère de l'incarnation de Denys le Petit', *Revue d'histoire ecclésiastique*, 44 (1949), pp. 463–87, here pp. 474–80: Bede and historians.

89. Abbo, *Computus vulgaris*, ed. Jacques-Paul Migne, *Patrologia Latina*, 90 (1854), col. 731: *calculator*; col. 758: alphabet; cols 953ff: foot-sundial; col. 823: third cycle (misread as 1615 instead of 1065). See Alfred Cordoliani, 'Les manuscrits de la bibliothèque de Berne provenant de l'abbaye de Fleury au XIe siècle. Le comput d'Abbon', *Zeitschrift für schweizerische Kirchengeschichte*, 52 (1958), pp. 135–50. For a discussion of the tables, see Alfred Cordoliani, 'Contribution à la littérature du comput ecclésiastique au moyen âge', *Studi medievali*, III/1 (1960), pp. 107–37, 169–208, here pp. 117–37, 169–173. For a study of Abbo's work as a whole, see Eva-Maria Engelen, *Abbo von Fleury. Philosophie und Wissenschaft im Zeichen der Zeit*, Diss. phil., Constance (1990); Borst (as n. 20), pp. 60–9.

90. 'Prologus', ed. José M. Millás Vallicrosa, *Assaig d'història de las idees físiques i matemàtiques a la Catalunya medieval*, 1 (1931), pp. 273ff: *computatio*. 'Sententie astrolabii', ibid., pp. 275, 280: *horologium*; pp. 281–4: *computare*; pp. 284–6: odd and even hours. 'De mensura astrolabii', ibid., p. 298: numerandi calculatio. On the arrangement of this corpus, see Werner Bergmann, *Innovationen im Quadrivium des 10. und 11. Jahrhunderts. Studien zur Einführung von Astrolab und Abakus im lateinischen Mittelalter* (1985), pp. 122–47; Guy Beaujouan, 'Les Apocryphes mathématiques de Gerbert', in: *Gerberto. Scienza, storia e mito*, ed. Michele Tosi (1985), pp. 645–58. My findings regarding the 'Constance Fragment' prove that the corpus was known in Fleury as early as *c.*995; see Borst (as n. 20), pp. 30–52, 112–27.

91. Gerbert of Aurillac, *De rationali et ratione uti*, c. 6, ed. Jacques-Paul Migne, *Patrologia Latina*, 139 (1853), cols 161ff: Aristotle, the heavens and the sun; c. 9, col. 164: numbers and time. See Carla Frova, 'Gerberto philosophus: il De rationali et ratione uti', in: *Gerberto* (as n. 90), pp. 351–77; Pierre Riché, *Gerbert d'Aurillac, le pape de l'an mil* (1987), pp. 181ff, 189–92.

92. Gerbert, *Regulae de numerorum abaci rationibus*, ed. Nicolaus Bubnov, *Gerberti opera mathematica* (1899), pp. 7–11: *digiti* and *articuli*, discussed in: *Commentarius in Gerberti regulas*, I, 2, 2, p. 252; Gerbert, *Geometria*, VI, 2, pp. 80ff: whole numbers and fractions; VI, 3, pp. 84ff: *abacista*. On Gerbert's controversial role in introducing the abacus, see Bergmann (as n. 90), pp. 185–215. Gerbert's collected letters, no. 183, *MGH Briefe der deutschen Kaiserzeit*, 2 (1966), p. 217: abacus-number; no. 153, pp. 180ff: *horologia*. See Landes (as n. 5), pp. 53ff, 64ff; Borst (as n. 20), pp. 52–6; an unconvincing contribution to the subject is made by

McCluskey (as n. 39), pp. 21ff. Murray (as n. 40), pp. 163–7, provides a stimulating discussion of the rationality of the abacus.

93. Gerbert (?), *Liber de astrolabio*, III, 3, ed. Bubnov (as n. 92), p. 126: *computare*, 'add up'; I, 1, p. 116: ecclesiastical purposes; II, 10, p. 122: *calculator*, 'pointer'; IX, p. 133: *computare*, 'to be read off'. On the author question, see Bergmann (as n. 90), pp. 148–63; Beaujouan (as n. 90), p. 651, with whom I agree regarding the minor deviations from Gerbert's own writings; see Borst (as n. 20), pp. 48, 78. On the pointer, see Willy Hartner, 'The Principle and Use of the Astrolabe', now in: Hartner *Oriens – Occidens. Ausgewählte Schriften zur Wissenschafts- und Kulturgeschichte*, I (1968), pp. 287–311, here pp. 300ff, 309ff. On the etymology of the pointer, see Paul Kunitzsch, *Glossar der arabischen Fachausdrücke in der mittelalterlichen europäischen Astrolabliteratur*, Nachrichten der Akademie der Wissenschaften in Göttingen, Phil. -hist. Kl. 1982/11 (1983), pp. 455–571, here pp. 538ff: the quoted definition and other references.

94. *De aggregatione naturalium numerorum*, ed. Maximilian Curtze, MS no. 14836 of the Königliche Hof- und Staatsbibliothek in Munich, *Zeitschrift für Mathematik und Physik*, 40 (1895), supplement pp. 75–142, here p. 106: *abacista*; p. 108: *compotiste*. This is amended in Borst (as n. 9), pp. 77ff. Since then I have discovered a second piece of textual evidence: Biblioteca Apostolica Vaticana, Codex Palatinus latinus 1356, 115r–116r, under the heading *Libellus abaci*.

95. Franco of Liège, *De quadratura circuli* I E, ed. Menso Folkerts and Alphons J. E. M. Smeur, 'A Treatise on the Squaring of the Circle by Franco of Liège of about 1050', *Archives internationales d'histoire des sciences*, 26 (1976), pp. 59–105, 225–53, here p. 67. See also Paul L. Butzer, 'Mathematics in the Region Aachen–Liège–Maastricht from Carolingian Times to the 19th Century', *Bulletin de la Société Royale de Sciences de Liège*, 51 (1982), pp. 5–30, here pp. 8–10.

96. Notker Labeo, *De quatuor questionibus computi*, ed. Paul Piper, *Nachträge zur älteren deutschen Literatur* (1898), pp. 312–18, here p. 313: *compotista*; p. 317: *calculator*. For an analysis which no other has succeeded in replacing, see Gabriel Meier, 'Die sieben freien Künste im Mittelalter', part 2, in: *Jahresbericht über die Lehr- und Erziehungsanstalt des Benediktinerstiftes Maria Einsiedeln im Studienjahr 1886/87* (1887), pp. 3–36, here pp. 11ff. Notker is overlooked by Alfred Cordoliani, 'L'évolution du comput ecclésiastique à Saint Gall du VIIIe au XIIe siècle', *Zeitschrift für schweizerische Kirchengeschichte*, 49 (1955), pp. 288–323. For a study of the effects on Ekkehard, see Borst (as n. 20), pp. 72ff.

97. *Die Werke Notker des Deutschen*, ed. James C. King and Petrus W. Tax, 9 (1981), p. 346. See also Grimm (as n. 62), vol. 12 (1885), cols 2427ff. The religious background is denied by Hans Kaletsch, *Tag und Nacht. Die Geschichte unseres Kalenders* (1970), p. 41.

98. *Musica Hermanni Contracti*, ed. Leonard Ellinwood (1952), p. 24: *unanimis omnium assertio et insuperabilis naturae veritas*; pp. 18ff: the week and sounds; p. 24: *structura*. Hans Oesch, *Berno und Hermann von Reichenau als Musiktheoretiker* (1961), pp. 228ff, undervalues this aspect; Arno Borst, 'Ein Forschungsbericht Hermanns des Lahmen', *Deutsches Archiv*, 40 (1984), pp. 379–477, here pp. 397ff; Borst (as n. 9), pp. 94ff,

158.

99. Hermannus Contractus, *Über das Astrolab*, c. 8, ed. Joseph Drecker, *Isis*, 16 (1931), pp. 200–19, here p. 211: on the astrolabe *ad astronomicam horologicamve disciplinam*; c. 5, p. 208: *calculator*, as above, n. 93; *De mensura horologii*, ed. Jacques-Paul Migne, *Patrologia Latina*, 143 (1853), cols 405–8: the column-sundial as a *horologicum instrumentum*. See Werner Bergmann, 'Der Traktat "De mensura astrolabii" des Hermann von Reichenau', *Francia*, 8 (1980), pp. 65–103, here pp. 69–75; Borst (as n. 20), pp. 77–82.

100. Hermannus Contractus, *Martyrologium*, Auszüge, ed. Dümmler (as n. 84), pp. 208–13. See Borst (as n. 98), pp. 398–406; John McCulloh, 'Hermann the Lame's Martyrology through Four Centuries of Scholarship', *Analecta Bollandiana*, 104 (1986), pp. 349–70.

101. On the unprinted 'Compotus', see Borst (as n. 98), pp. 427–31, the quotation p. 428; see also, despite its lack of new findings, Werner Bergmann, 'Chronographie und Komputistik bei Hermann von Reichenau', in: *Historiographia mediaevalis. Studien zur Geschichtsschreibung und Quellenkunde des Mittelalters. Festschrift für Franz-Josef Schmale* (1988), pp. 103–17.

102. Hermannus Contractus,*Chronicon*, a. 456, *MGH Scriptores*, 5 (1844), p. 83 and a. 550, p. 88 on the incorrect calculation of Easter and its correction. See Arno Borst, 'Hermann der Lahme und die Geschichte', in: Borst (as n. 61), pp. 135–54. On doubts at the end of his life, in the unprinted *Prognostica*, Borst (as n. 98), pp. 436–40.

103. 'Necrologium Benedictoburanum', *MGH Necrologia Germaniae*, 1 (1888), p. 4 on 13 March, with the misreading *1147*. A correction is offered by Hartmut Hoffmann, *Buchkunst und Königtum im ottonischen und frühsalischen Reich*, 1 (*Schriften der MGH*, 30/1, 1986), pp. 431ff. On similar entries of the writer's dates in a computistical manuscript from Illmünster, see Borst (as n. 9), pp. 298ff.

104. Gerlandus, *Regulae super abacum*, ed. Peter Treutlein, *Scritti inediti relativi al calcolo dell' abaco, Bollettino di bibliografia e di storia delle scienze matematiche e fisiche*, 10 (1877), pp. 595–647, here pp. 595–607, refers several times to *abacistae*. See Alfred Cordoliani, 'Notes sur un auteur peu connu: Gerland de Besançon', *Revue du moyen âge latin*, 1 (1945), pp. 411–19, here pp. 417–19; Borst (as n. 9), pp. 111ff. On the computus, see Alfred Cordoliani, 'Le Comput de Gerland de Besançon', *Revue du moyen âge latin*, 2 (1946), pp. 309–13, here p. 311 on the martyrology. Extracts from the computus ed. Cordoliani (as n. 88), pp. 484–7, here p. 484: *calculatores*. On the dating system, see Borst (as n. 98), pp. 465ff.

105. *Regulae domni Oddonis super abacum*, ed. Martin Gerbert, *Scriptores ecclesiastici de musica sacra potissimum*, 1 (1784), p. 296; also ed. Jacques-Paul Migne, *Patrologia Latina*, 133 (1854), col. 807. On the author question, see Borst (as n. 9), pp. 116–18.

106. Honorius Augustodunensis, *Elucidarium*, I, 19, ed. Yves Lefèvre, *L'Elucidarium et les lucidaires* (1954), p. 364: the Creation; I, 36, p. 367: Satan; I, 90, p. 377: Adam; I, 128, p. 384: Christ's birth; III, 50, p. 457: Last Judgement; I, 156–7, p. 389: forty hours. See Jacques Le Goff, *Medieval Civilization, 400–1500* (1988), pp. 165ff. On shortage of time, see Murray (as n. 40), pp. 105–7.

107. Hugh of Saint-Victor, *De tribus maximis circumstantiis gestorum*, ed. William M. Green, *Speculum*, 18 (1943), pp. 484–93, here pp. 489–91. See Joachim Ehlers, *Hugo von St. Viktor. Studien zum Geschichtsdenken und zur Geschichtsschreibung des 12. Jahrhunderts* (1973), pp. 136–55; Borst (as n. 9), pp. 180ff; John B. Friedman, 'Les Images mnémotechniques dans les manuscrits de l'époque gothique', in: *Jeux de mémoire* (as n. 75), pp. 169–84, here pp. 173ff. On the astrolabe, see Borst (as n. 20), pp. 87ff. On the book metaphor, see Ernst Robert Curtius, *European Literature and the Latin Middle Ages* (1978), pp. 311–19; Hans Blumenberg, *Die Lesbarkeit der Welt* (1983), pp. 51–3, poorly informed about Hugh's learning.

108. Guido Augiensis, *Regulae de arte musica*, c. 1, ed. Edmond de Coussemaker, *Scriptorum de musica medii aevi nova series*, 2 (1867), p. 152. Here I am amending the author's name, following Michael Bernhard, 'Das musikalische Fachschrifttum im lateinischen Mittelalter', in: *Geschichte der Musiktheorie*, 3 (as n. 36), pp. 37–103, here p. 59.

109. Marianus Scottus, *Chronicon*, a. 1050–1091, *MGH Scriptores*, 5 (1844), pp. 556–60: autobiography; a. 548, p. 538: Dionysius; a. 700, p. 544 and a. 747, p. 546: Bede as *compotator*. On the work, see Anna-Dorothee von den Brincken, 'Marianus Scottus. Unter besonderer Berücksichtigung der nicht veröffentlichten Teile seiner Chronik', *Deutsches Archiv*, 17 (1961), pp. 191–238.

110. Bernold, *Chronicon, De regularibus patrum*, *MGH Scriptores*, 5, p. 393: Hermann as a *compotistarum subtilissimus*; a. 1093, p. 457: as an *egregius calculator*; cf. below, n. 116. See Borst (as n. 98), pp. 461–3. On the calendar, see Rolf Kuithan and Joachim Wollasch, 'Der Kalender des Chronisten Bernold', *Deutsches Archiv*, 40 (1984), pp. 478–531.

111. Frutolf, *Chronica*, a. 1093, *Freiherr vom Stein-Gedächtnisausgabe*, 15 (1972), p. 106, specifies the hour of a solar eclipse; p. 108, in the case of a lunar eclipse, the lunar phase of 14 days according to the full moon. See Borst (as n. 98), pp. 463ff. On Frutolf's circle, see Otto Meyer 'Weltchronistik und Computus im hochmittelalterlichen Bamberg', now in: Meyer, *Varia Franconiae Historica*, 2 (1981), pp. 768–87, here p. 769: *compotistae nostri temporis*, in the case of Frutolf's pupil, Heimo. For him, see Anna-Dorothee von den Brincken, 'Die Welt- und Inkarnationsära bei Heimo von St. Jakob', *Deutsches Archiv*, 16 (1960), pp. 155–94, here p. 183: *nostri cronografi et compotiste*.

112. Pseudo-Bede, *De mundi celestis terrestrique constitutione*, I, 50, ed. Charles Burnett (1985), p. 22, also I, 317–19, pp. 44–6. The distinction, unthinkable in the ninth and tenth centuries, confirms the dating system used by the editor, pp. 1–3, against Niermeyer (above, n. 72).

113. *Gesta Treverorum*, c. 21, *MGH Scriptores*, 8 (1848), p. 195. On Joshuah, see Alfred Haverkamp, 'Die Juden im mittelalterlichen Trier', *Kurtrierisches Jahrbuch*, 19 (1979), pp. 5–57, here pp. 27ff.

114. Sigebert of Gembloux, *Chronicon*, a. 532, *MGH Scriptores*, 6 (1844), p. 316: criticism of Dionysius; similarly, a. 979, p. 352; a. 1076, p. 363. *Liber decennalis*, III, 60, *MGH Quellen zur Geistesgeschichte*, 12 (1986), pp. 284ff: Dionysius and moderns; II, 64–71, pp. 252–6: astuteness and truth; III, 7, pp. 258ff: Marianus. See Borst (as n. 98), pp. 464ff; Joachim Wiesenbach, *MGH Quellen zur Geistesgeschichte*, 12, pp. 9–168.

115. Adam of Bremen, *Gesta Hammaburgensis ecclesiae pontificum*, III, 66,

MGH Scriptores rerum Germanicarum, 2 (1917), p. 213: *calculare*; I, 35, p. 38; I, 45, p. 46: *compotus*. On the work, see Franz-Josef Schmale, 'Adam von Bremen', in: *Die deutsche Literatur des Mittelalters. Verfasserlexikon*, 1 (1978), cols 50–4.

116. William of Hirsau, *Statuta Hirsaugensia*, II, 34, ed. Jacques-Paul Migne, *Patrologia Latina*, 150 (1854), col. 1089: liturgical time; II, 44, col. 1104: corn harvest; Bernold, *Chronicon*, a. 1091, *MGH Scriptores*, 5, p. 451: *horologium* and *compotus*. See also Landes (as n. 5), pp. 69ff; Borst (as n. 98), pp. 460ff. On the astrolabe Borst (as n. 20), pp. 82ff; Joachim Wiesenbach, 'Wilhelm von Hirsau, Astrolab und Astronomie im 11. Jahrhundert', in Klaus Schreiner (ed.), *Hirsau St. Peter und Paul 1091–1991* (1991), pp. 109–54. The roughly simultaneous northern French *Horologium stellare monasticum*, ed. Giles Constable, in: *Corpus consuetudinum monasticarum*, 6 (1975), pp. 1–18, centres solely on the stars and their position in relation to individual monastery buildings.

117. Elias Steinmeyer and Eduard Sievers, *Die althochdeutschen Glossen*, 3 (1969), p. 655, after the Munich MS 14689 from St Emmeram. On its dating, see Borst (as n. 9), p. 138.

118. William of Malmesbury, *Gesta regum Anglorum*, II, 118, *Rerum Britannicarum scriptores*, 90/1 (1887), p. 122: *sine computo*. The old meaning *lunaris compotus* in William of Malmesbury, *Gesta pontificum Anglorum*, IV, 164, ibid., vol. 52 (1870), p. 300. On the astrolabe, see Borst (as n. 20), p. 86. On the alliance between the monarchy and arithmetic, see Murray (as n. 40), pp. 180ff, 194–203.

119. *Leges Edwardi Confessoris*, c. 32, B 13, ed. Felix Liebermann, *Die Gesetze der Angelsachsen*, 1 (1903), p. 657.

120. *Charta Henrici I regis*, quoted from the corrected proofs of Henry G. Richardson, 'Henry I's Charter to London', *English Historical Review*, 42 (1927), pp. 80–7, here p. 82. An interpretation which is too technical is offered by Liebermann (as n. 119), vol. 3 (1916), p. 304, *ad compotum*: thus the king promised merely to count the coins, and not to have them thoroughly checked. For a summary of the formation of the Exchequer, see Gerald L. Harriss, 'Exchequer', in: *Lexikon des Mittelalters*, 4 (1989), cols 156–9.

121. Richard Fitz Nigel, *Dialogus de scaccario*, I, 5, ed. Charles Johnson (1983), pp. 24–6: *calculator*; I, 3, p. 11: *computatores*; I, 14, p. 62: *annales compotorum*. Further evidence of this usage of the word is provided by the *Dictionary of Medieval Latin from British Sources*, ed. Ronald E. Latham, 2 (1981), pp. 414ff. About 1150 Adelard of Bath called the pointer on the astrolabe *computator*, a term which scarcely caught on: *Libellus de opere astrolapsus*, ed. Bruce G. Dickey, *Adelard of Bath. An Examination based on Heretofore Unexamined Manuscripts*, Diss. phil. Toronto (1982), p. 200. See Kunitzsch (as n. 93), pp. 538ff and below, n. 170.

122. Jacques Le Goff, 'Au moyen âge: temps de l'église et temps du marchand', *Annales ESC*, 15 (1960), pp. 417–33; Le Goff, *Your Money or your Life: Economy and Religion in the Middle Ages* (1988), pp. 33–45.

123. Peter Abelard, *Dialectica*, I, 2, 2, 2–3, ed. Lambertus M. De Rijk (1970), pp. 61–5. See Arno Borst, 'Historical Time in the Writings of Abelard', in: Borst (as n. 61), pp. 72–88.

124. Abelard, *Dialectica* I, 2, 2, 2, p. 61; I, 2, 3, 2, p. 78; I, 2, 3, 18, p. 108.

125. Ibid., I, 2, 2, 1, p. 59: arithmetic; II, 2, 9, pp. 216ff: astronomy; IV, 1, 'Prologus', p. 469: *mathematica*; I, 2, 3, 10, pp. 99ff: geometry. See Borst (as n. 9), pp. 212ff. A more positive view of arithmetic is given in the early work *Theologia summi boni*, I, 6, ed. Ursula Niggli, Philosophische Bibliothek, 395 (1988), pp. 44–7. On the devaluing of nature and natural science, see Borst (as n. 20), pp. 85ff for the quotation.

126. Abelard, *Historia calamitatum*, ed. Jacques Monfrin (1967), pp. 70, 81, 94. On Abelard's lack of interest in numbers and dates, see Murray (as n. 40), p. 177.

127. Otto of Freising, *Chronica sive Historia de duabus civitatibus, Epistola*, *MGH Scriptores rerum Germanicarum*, 45 (1912), p. 5: *cronographi*; V, 18, p. 248: Bede; I, 5, p. 43: *annorum supputatio*. John of Salisbury, *Historia pontificalis, Prologus*, ed. Marjorie Chibnall (1956), pp. 2ff: Bede, *series temporum* and *cronici scriptores*. On the dates and numbers used by these historians, see Murray (as n. 40), pp. 174–80. On the historicization of historiography, see Bernard Guenée, *Les premiers pas de l'histoire de l'historiographie en Occident au XIIe siècle*, Académie des inscriptions et belles lettres, Comptes rendus (1983), pp. 136–52.

128. *Geschichtsschreibung und Geschichtsbewusstsein im Spätmittelalter*, ed. Hans Patze, *Vorträge und Forschungen*, 31 (1987), pp. 81, 551, shows that the connection between historiography and computation did not completely break, but became more tenuous.

129. *Decretum Gratiani*, D. 38, c. 5, ed. Emil Friedberg, *Corpus iuris canonici*, 1 (1879), cols 141ff: *computus*, supposedly according to Augustine, but in fact according to Haito of Basle (as above, n. 68); D. 37, c. 10, col. 138: arithmetic, geometry, music. On the second passage, see Murray (as n. 40), p. 178.

130. See also Adolf P. Juschkewitsch, *Geschichte der Mathematik im Mittelalter* (1964), pp. 175–325, 349–57; W. Montgomery Watt, *The Influence of Islam on Medieval Europe* (1972). On the reservations of canon law, see Borst (as n. 20), pp. 93ff.

131. Alfred Nagl, 'Über eine Algorismus-Schrift des XII. Jahrhunderts und über die Verbreitung der indisch-arabischen Rechenkunst und Zahlzeichen im christlichen Abendlande', *Zeitschrift für Mathematik und Physik*, 34 (1889), Historisch-literarische Abteilung, pp. 129–46, 161–70. On the Viennese manuscript finally, Otto Mazal and Eva Irblich, *Wissenschaft im Mittelalter* (1978), p. 190, no. 161. On the algorithm, see Karl Menninger, *Zahlwort und Ziffer. Eine Kulturgeschichte der Zahl*, 2 (1979), pp. 225–7, with illustration p. 239. Murray (as n. 40), pp. 167–74 is familiar with the sources, but when he deals with the High Middle Ages he loses interest in the computus. Landes (as n. 5), p. 78, makes the townsmen into pioneers of the new arithmetic. Hans Patze rightly emphasizes how belatedly they adapted themselves to it: Hans Patze, 'Neue Typen des Geschäftsschriftgutes im 14. Jahrhundert', in: *Der deutsche Territorialstaat im 14. Jahrhundert*, ed. Hans Patze, *Vorträge und Forschungen*, 13/1 (1970), pp. 9–64, here p. 64.

132. Reiner of Paderborn, *Compotus emendatus*, II, 1–4, ed. Walter E. van Wijk, *Le comput emendé de Reinherus de Paderborn, Verhandelingen der K. Nederlandse Akademie*, N.F. 57/3 (1951), pp. 48–50: arithmetical errors,

Hermann's lunar month; I, 24, pp. 44–6: the moon's course; I, 12, p. 28: Moses; II, 4, p. 50: age of the Julian calendar; II, 7, p. 56: the Creation; I, 1, p. 16: *compotus*; II, 8–15, pp. 56–70: the Resurrection of Christ; 'Praefatio', p. 10: the Church and reputation. On the actual age of the Jewish calendar, see above, n. 29. A closer study needs to be made of Reiner's educational career; Reiner's name is even missing in Peter Classen, *Studium und Gesellschaft im Mittelalter, Schriften der MGH*, 29 (1983), as well as in *Schulen und Studium im sozialen Wandel des hohen und späten Mittelalters*, ed. Johannes Fried, *Vorträge und Forschungen*, 30 (1986).

133. Master Gregorius, *Narracio de mirabilibus urbis Romae*, c. 12, ed. Robert B. C. Huygens, *Textus minores*, 42 (1970), pp. 20ff: Dioscuri; c. 29, pp. 28ff: Caesar's tomb. See Gerd Tellenbach, 'Die Stadt Rom in der Sicht ausländischer Zeitgenossen (800–1200)', *Saeculum*, 24 (1973), pp. 1–40, here pp. 10, 35–7; John Osborne, *Master Gregorius, The Marvels of Rome* (1987), pp. 60, 88–94. The older tradition: *Mirabilia urbis Rome*, c. 26, ed. Percy Ernst Schramm, *Kaiser, Rom und Renovatio*, 2 (1929), pp. 88ff: Dioscuri; c. 13, p. 80: Caesar's tomb.

134. Alexander de Villa Dei, *Massa compoti*, 'Praefatio', ed. Walter E. van Wijk, *Le Nombre d'or. Étude de chronologie technique* (1936), p. 52: time-fixing, Caesar. Reinhard Elze has pointed out to me that the examplary year, 1200, given in vv. 281ff, p. 59, is not necessarily the year of writing, as Wijk believes (p. 31). On the mnemonic verses and their influence, see Bernhard Bischoff, 'Ostertagtexte und Intervalltafeln', in: Bischoff (as n. 74), pp. 192–227. On the *computus minor*, see above, n. 82.

135. John de Sacro Busto (*sic*), *Libellus de anni ratione seu ut vocatur vulgo Computus ecclesiasticus*, ed. Philipp Melanchthon (reprinted 1558), fo. I 6v: definition; fo. M 8r: leap-day; fo. O 1r: Council. On the author, most recently, Francis B. Brévart and Menso Folkerts, 'Johannes de Sacrobosco', in: *Die deutsche Literatur des Mittelalters. Verfasserlexikon*, 4 (1983), cols 731–6; on the computus, see Borst (as n. 98), p. 468; on the place of astronomy and the astrolabe in the university see Borst (as n. 20), pp. 94–6.

136. Vincent of Beauvais, *Speculum doctrinale*, XVI, 9 (1624), col. 1509. Earlier, at XVI, 6, col. 1507, he repeated verbatim Isidore's praise of the *computus*, adopted from Cassiodorus (see above, n. 47). Justification for his method of dating: *Apologia actoris*, c. 5, ed. Anna-Dorothee von den Brincken, 'Geschichtsbetrachtung bei Vincenz von Beauvais', *Deutsches Archiv*, 34 (1978), pp. 410–99, here p. 471.

137. Albertus Magnus, *Summa theologica*, II, 11, 59, in: *Opera omnia*, ed. Auguste Borgnet, 32 (1895), p. 586: *in computo ecclesiastico*. Thomas Aquinas, *In quattuor libros Sententiarum*, IV, 13, 1, 2d, in: *Opera omnia*, ed. Roberto Busa, 1 (1980), p. 491: *secundum ecclesiae computum*. See Anneliese Maier, 'Die Subjektivierung der Zeit in der scholastischen Philosophie', *Philosophia naturalis*, 1 (1950/2), pp. 361–98, here pp. 369–71, 376–9; on the assessment of the astrolabe, see Borst (as n. 20), pp. 96ff. Michael Bernhard has indicated to me that there is a marginal exception: in the late thirteenth century Parisian music theorists adhered to the Aristotelian definition of time as number. References in Ulrich Michels, *Die Musiktraktate des Johannes de Muris* (1970), p. 72. For

an account of the consequences, see below, n. 157.

138. Roger Bacon, *Compotus*, 'Prologus', in: *Opera hactenus inedita*, ed. Robert Steele, 6 (1926), pp. 2ff. Bacon used the definition of Robert Grosseteste, *Compotus factus ad correctionem communis kalendarii nostri*, c. 1, ibid., p. 213. On Grosseteste, see Borst (as n. 98), pp. 468ff; Richard C. Dales, 'The Computistical Works ascribed to Robert Grosseteste', *Isis*, 80 (1989), pp. 74–9. On the three types, see above, n. 58.

139. Bacon, *Compotus*, II, 18–19, pp. 146–50: criticism of the calendar; II, 1, pp. 87–9: *compotiste*. See Borst (as n. 98), pp. 470ff. On Bacon's judgement of the astrolabe, see Borst (as n. 20), pp. 97ff. On Islamic water-clocks, see Eilhard Wiedemann and Fritz Hauser, *Über die Uhren im Bereich der islamischen Kultur*, Nova Acta Leopoldina, 100/5 (1915); Donald R. Hill, *Arabic Water-Clocks* (1981); on sundials, see Karl Schoy, 'Gnomonik der Araber', *Die Geschichte der Zeitmessung und der Uhren*, ed. Ernst von Bassermann-Jordan, 1 F, 1925; René R. J. Rohr, *Die Sonnenuhr, Geschichte, Theorie, Funktion* (1982), pp. 168–79.

140. Roger Bacon, *Opus maius*, IV, ed. John H. Bridges, 1 (1897), pp. 187–210, 269–85. The assertion made by Crombie (as n. 57), p. 53, that Bacon's proposals for reform were put into practice in 1582, ignores the fundamental change in the conception of time and numbers that occurred between the thirteenth and sixteenth centuries.

141. Gulielmus Durandus, *Rationale divinorum officiorum*, VIII, 1 (1568), fo. 466r: definition; VIII, 10, fo. 478v: errors, also VIII, 11, fo. 479r; dating of 1286 according to VIII, 9, fo. 477v. Specialist research overlooks the computistical part; Georg Steer, 'Durandus', in: *Die deutsche Literatur des Mittelalters. Verfasserlexikon*, 2 (1980), cols 245–7, at least refers to it.

142. Lynn Thorndike, 'Computus', *Speculum*, 29 (1954), pp. 223–38, brings together works from the fourteenth and fifteenth centuries still bearing the title 'computus', mostly with a popularizing additional term (pp. 224–7). To the list, for example, we would have to add the text of the Constance schoolmaster Burkard Fry, *Instructio parochorum de computo ecclesiastico* (1436). More problem-conscious authors chose different titles.

143. John of Montpellier, 'Tractatus quadrantis', ed. Nan L. Hahn, *Medieval Mensuration: Quadrans vetus and Geometrie due sunt partes principales*, Transactions of the American Philosophical Society, 82/8 (1982), pp. 6–113. On the history of the solar quadrant, see ibid., pp. xxxii–xxxvi, with older literature; there is to be added Ernst Zinner, *Deutsche und niederländische astronomische Instrumente des 11.–18. Jahrhunderts* (1967), pp. 154–63. For a study of the remaining errors, see Margarida Archinard, 'The Diagram of Unequal Hours', *Annals of Science*, 47 (1990), pp. 173–90.

144. Robertus Anglicus, *Commentarius in Sphaeram*, c. 11, ed. Lynn Thorndike, *The Sphere of Sacrobosco and Its Commentators* (1949), pp.179ff. For the most accurate account, see Lynn Thorndike, 'Invention of the Mechanical Clock about 1271 A.D.', *Speculum*, 16 (1941), pp. 242ff; Lynn White, *Medieval Technology and Social Change* (1962), pp. 119–29, with inappropriate conclusions about the history of the astrolabe; it is excluded by Donald R. Hill in *A History of Engineering in Classical and Medieval Times* (1984), pp. 241–5. A more relevant

examination is provided by Jenzen (as n. 59), pp. 15–28, who by studying Robert's sentences, should have been warned against drawing false theological conclusions.

145. Brunetto Latini, *Li livre dou tresor*, I, 118, 3, ed. Francis J. Carmody (1975), p. 103: *conteour*; I, 118, 6, p. 104: *li contes*. Bono Giamboni directly translated the passage using the word *conto*. For further references, see Salvatore Battaglia, *Grande dizionario della lingua italiana*, 3 (1964), pp. 436ff, 660–4. On the heightened numerical consciousness of historiography, see Murray (as n. 40), pp. 180–7; on trade, pp. 189–94.

146. Dante Alighieri, *Il fiore*, VIII, 3, in: *Opere minori*, ed. Domenico De Robertis and Gianfranco Contini, 1/1 (1984), p. 572: reckoning; CLIV, 5, p. 720: balancing. See *Enciclopedia Dantesca*, ed. Umberto Bosco, 2 (1970), p. 178.

147. Wilhelm Meyer-Lübke, *Romanisches etymologisches Wörterbuch* (1935), p. 199, on the word. On the object, see Menninger (as n. 131), pp. 245ff.

148. Peter Herde, *Beiträge zum päpstlichen Kanzlei- und Urkundenwesen im dreizehnten Jahrhundert* (1967), pp. 181–90. On the effects, see below, n. 193.

149. Ducange (as n. 28), cols 473ff on the word. For a concise study of the object, see Elisabeth Lalou, 'Chambre des comptes', in: *Lexikon des Mittelalters*, 2 (1983), cols 1673–5.

150. Grimm (as n. 62), vol. 11 (1873), col. 1743 on the word. On the object, see Menninger (as n. 131), p. 161.

151. *MGH Constitutiones et acta publica imperatorum et regum*, 1 (1893), p. 487, no. 341: the emperor's decree of 1191. See Wilhelm Erben, *Die Kaiser- und Königsurkunden des Mittelalters in Deutschland, Frankreich und Italien* (1907), pp. 324–6. *Corpus der altdeutschen Originalurkunden bis zum Jahr 1300*, ed. Friedrich Wilhelm, 1 (1932), pp. 46–53, no. 26: Lucerne decree of 1252.

152. Rolf M. Kully, 'Cisiojanus. Comment savoir le calendrier par coeur', in: *Jeux de mémoire* (as n. 75), pp. 149–56, on the Latin version. On the German versions, see Arne Holtorf, 'Cisioianus', in: *Die deutsche Literatur des Mittelalters. Verfasserlexikon*, 1 (1978), cols 1285–9.

153. Arnold Esch, 'Zeitalter und Menschenalter. Die Perspektiven historischer Periodisierung', *Historische Zeitschrift*, 239 (1984), pp. 309–51, here pp. 333–49.

154. Alexander of Roes, 'Noticia seculi', cc. 6–7, *MGH Staatsschriften des späteren Mittelalters*, 1/1 (1958), pp. 151ff; cc. 20–3, pp. 167–71. See Herbert Grundmann, 'Über die Schriften des Alexander von Roes', now in: Grundmann (as n. 7), pp. 196–274, here pp. 202–6; Bernhard Töpfer, *Das kommende Reich des Friedens. Zur Entwicklung chiliastischer Zukunftshoffnungen im Hochmittelalter* (1964), pp. 45ff, 146ff. On the apocalyptic mood of 1240–60, see Hans Martin Schaller, 'Endzeit-Erwartung und Antichrist-Vorstellungen in der Politik des 13. Jahrhunderts', in: *Festschrift für Hermann Heimpel*, 2 (*Veröffentlichungen des Max-Planck-Instituts für Geschichte*, 36/2, 1972), pp. 924–47.

155. Elias Salomon, 'Scientia artis musicae', c. 17, ed. Gerbert (as n. 105), vol. 3 (1784), p. 36. Paul Lehmann is not familiar with this document: Paul Lehmann, 'Blätter, Seiten, Spalten, Zeilen', in: Paul Lehmann (as n. 42), vol. 3 (1960), pp. 1–59, here pp. 17–25. On the resistance to Indian

numerals, see Menninger (as n. 131), pp. 244ff; Murray (as n. 40), pp. 169–72.

156.   *Exafrenon pronosticacionum temporis*, c. 1, ed. John D. North, *Richard of Wallingford. An Edition of his Writings with Introduction, English Translation and Commentary*, 1 (1976), pp. 184–92. In the late fourteenth century *compotestas* was still being translated as the semi-French *countoures*, ibid., p. 187. On the work, ibid., vol. 2 (1976), pp. 83–97, 100–2. The theory of despondency in the fourteenth century is put forward by Hans Blumenberg, *Der Prozess der theoretischen Neugierde* (1980), pp. 151–7.

157.   Bernhard Schimmelpfennig, 'Heiliges Jahr', in: *Lexikon des Mittelalters*, 4 (1989), cols 2024ff: on Boniface and Clement. Jean de Meurs, *Notitia artis musicae*, II, 1, ed. Ulrich Michels, *Corpus scriptorum de musica*, 17 (1972), p. 65: the measure of movement. *Arithmetica speculativa*, I, 1, ed. Hubertus L. L. Bussard, *Die 'Arithmetica speculativa' des Johannes de Muris. Scientiarum Historia*, 13 (1971), pp. 103–32, here p. 116: distinguishing numbers. On his life and works, including astronomical and astrological works, see Michels (as n. 137), pp. 1–15; on time and numbers, pp. 69–75. On the preliminary stages of music theory, see above, n. 137.

158.   Ferdinand Kaltenbrunner, *Die Vorgeschichte des Gregorianischen Kalenderreform, Sitzungsberichte der Österreichischen Akademie der Wissenschaften*. Phil. -hist. Kl. 82/3 (1876), pp. 289–414, here pp. 315–22 on the field of the *Epistola*. Its text is edited and assessed by Christine Gack, *Johannes de Muris, Super reformatione antiqui kalendarii*, Diss. phil., Tübingen (1990).

159.   *Regule seu proprie canones qui dicuntur Pheffer Kuchel*, ed. Thorndike (as n. 142), pp. 234–8; the quotation, p. 238.

160.   For a summary, see Anneliese Maier, '"Ergebnisse" der spätscholastischen Naturphilosophie', now in: Anneliese Maier, *Ausgehendes Mittelalter. Gesammelte Aufsätze zur Geistesgeschichte des 14. Jahrhunderts*, 1 (1964), pp. 425–57; on details of the theory of movement, see Edward Grant, *Studies in Medieval Science and Natural Philosophy* (1981). A Middle English list of *calculatours*, ed. North (as n. 156), vol. 3 (1976), pp. 140ff, included late fourteenth-century astronomers and astrologers.

161.   White (as n. 144), pp. 123–5; Jean Gimpel, *The Medieval Machine: The Industrial Revolution of the Middle Ages* (1988), pp. 147–68, both with too much faith in technology. On the intellectual changes, see Jean Leclercq, 'Zeiterfahrung und Zeitbegriff im Spätmittelalter', in A. Zimmermann (ed.), *Antiqui und moderni. Traditionsbewusstsein und Fortschrittsbewusstsein im späten Mittelalter* (*Miscellanea Mediaevalia*, 9, 1974), pp. 1–20. On the social historical context, see Ferdinand Seibt, 'Die Zeit als Kategorie der Geschichte und als Kondition des historischen Sinns', in: *Die Zeit* (as n. 24), pp. 145–88, here pp. 164–73. On the physical preconditions, see Janich (as n. 16), pp. 228–45. For a comprehensive study which is nevertheless focused on the early modern period, see Landes (as n. 5), pp. 70–82.

162.   Jenzen (as n. 59), pp. 11–36 on the birth of the mechanical clock out of the astrolabe and the water-clock; Klaus Maurice, *Die deutsche Räderuhr*.

Zur Kunst und Technik des mechanischen Zeitmessers im deutschen Sprachraum, 2 (1976), pp. 16ff, no. 34, on the Nuremberg clock; ibid., vol. 1 (1976), pp. 33ff on the division of the hours there. On Bede, see above, n. 59. On the word 'Uhr', see Grimm (as n. 62), vol. 23 (1936), cols 731–8.

163. Hermann Heimpel, *Der Mensch in seiner Gegenwart. Acht historische Essays* (1957), pp. 11, 49ff, 59ff: non-synchrony and differences in the rhythms of clocks. On the clock metaphor, see Otto Mayr, 'Die Uhr als Symbol für Ordnung, Autorität und Determinismus', in: *Die Welt als Uhr. Deutsche Uhren und Automaten 1550–1650*, ed. Klaus Maurice and Otto Mayr (1980), pp. 1–9, here pp. 2ff. On the book metaphor, see above, n. 107.

164. Heinrich Seuse, *Horologium sapientiae*, 'Prologus', ed. Pius Künzle (1977), pp. 364ff. See also the Introduction, ibid., pp. 55–71; Leclercq (as n. 161), pp. 17–19.

165. Reiner Dieckhoff, 'Antiqui – moderni. Zeitbewusstsein und Naturerfahrung im 14. Jahrhundert', in: *Die Parler und der schöne Stil 1350–1400. Europäische Kunst unter den Luxemburgern*, ed. Anton Legner, 3 (1978), pp. 67–93, here pp. 67ff: Lorenzetti. For an informed study of this interpretation, see Ernst Jünger, *Das Sanduhrbuch* (1954), pp. 119–92; for a one-sidedly nautical view, see Robert T. Balmer, 'The Operation of Sand Clocks and their Medieval Development', *Technology and Culture*, 19 (1978), pp. 615–32. On Isidore, see above, n. 48.

166. Richard of Wallingford, *Tractatus horologici astronomici*, ed. North (as n. 156), vol. 1, pp. 444–523: the clock. See ibid., vol. 2, pp. 315–20, 361–70; Landes (as n. 5), pp. 83ff. *Declaraciones super kalendarium regine*, ed. North, vol. 1, pp. 558–63: horoscope. See vol. 2, pp. 371–8. For a summary, see North, 'Astrologie', in: *Lexikon des Mittelalters*, 1 (1980), cols 1141–3, and North, *Horoscopes and History*, *Warburg Institute Surveys*, 13 (1986).

167. Nicholas Oresme, *Le Livre du ciel et du monde*, II, 2, ed. Albert D. Menut and Alexander J. Denomy (1968), p. 282: regularity; p. 288: motion. See also Heribert M. Nobis, 'Astrarium', in: *Lexikon des Mittelalters*, 1 (1980), cols 1134ff.

168. Jacques Le Goff, *Time, Work and Culture in the Middle Ages* (1980), pp. 29–42. The example of Frankfurt am Main illustrates how a public mechanical clock regulated town life without a royal decree, in Jenzen (as n. 59), pp. 37–66. Le Goff's approach is further differentiated by Gerhard Dohrn-van Rossum, 'Zeit der Kirche, Zeit der Händler, Zeit der Städte', in: *Zerstörung und Wiederaneignung von Zeit*, ed. Rainer Zoll (1988), pp. 89–119, and by Dohrn-van Rossum, *Die Geschichte der Stunde. Uhren und moderne Zeitordnung* (1992), pp. 121–63.

169. *Exhortatio ad concilium generale Constantiense super correctione calendarii*, c. 1, ed. Giovanni Domenico Mansi, *Sacrorum conciliorum nova et amplissima collectio*, 28 (1785), col. 371: pennies and pounds; c. 3, col. 374: *praecisa veritas*; c. 2, cols 372ff: computation and astronomy; c. 6, col. 380: duration of the year and the Hebrews. See Kaltenbrunner (as n. 158), pp. 326–36. Modern researchers are uninterested in the relationship of the conciles to the computus; see the survey undertaken

by the editor in: *Das Konstanzer Konzil*, ed. Remigius Bäumer (*Wege der Forschung* 415, 1977), pp. 3–34.

170. Geoffrey Chaucer, *A Treatise on the Astrolabe*, I, 21, in: *The Works*, ed. Fred N. Robinson (1957), p. 549: *to calcule* and *calculer* (the *calculator* from above, nn. 93 and 99); 'Prologue', p. 546: astrologer and tables; I, 11, p. 547: calendar. Astrology, tables and *to calcule* are also mentioned in *The Canterbury Tales*, V, 1261–84, p. 141. *The Romaunt of the Rose*, B, 5026, p. 612: *compte*. On Chaucer's astrolabe, see Borst (as n. 20), pp. 100ff.

171. *The Booke of the Pylgrymage of the Sowle*, V, 1, ed. Katherine I. Cust (1859), pp. 73ff.: *this compute*, the equation of *seculum* and century, as the work of the *competister in the craft of kalendar*. On the author question, see ibid., p. iv; Walter F. Schirmer, *John Lydgate. Ein Kulturbild aus dem 15. Jahrhundert* (1952), p. 104. On *computacioun*: *Middle English Dictionary*, ed. Hans Kurath et al., 2/1 (1959), p. 478.

172. Nicholas of Cusa, *Die Kalenderverbesserung. De correctione kalendarii*, c. 2, ed. Viktor Stegemann and Bernhard Bischoff (1955), p. 14: *punctalis veritas*; p. 18: human understanding; c. 10, p. 84: discrepancy; p. 86: earlier regularities; c. 9, p. 80: astronomers; c. 3, p. 22: computists; c. 7, p. 56: wrong fixing; c. 8, pp. 68–72: reform proposals; c. 9, pp. 76–80: Hebrews and Greeks; c. 10, pp. 86–8: objections. See ibid., pp. xiii–lxxviii; Erich Meuthen, *Nikolaus von Kues 1401–1464. Skizze einer Biographie* (1985), pp. 39ff. As Zemanek (as n. 8), pp. 30, 32, shows, the Byzantine Orthodox calendar is in fact the most accurate of those still in use today.

173. Rudolf Klug, 'Johannes von Gmunden, der Begründer der Himmelskunde auf deutschem Boden', *Sitzungsberichte der Österreichischen Akademie der Wissenschaften*, Phil.-hist. Kl. 222/4 (1943), pp. 71–85; Konradin Ferrari d'Occhieppo, 'Die Osterberechnung als Kalenderproblem von der Antike bis Regiomontanus', in: *Regiomontanus-Studien*, ed. Günther Hamann, *Sitzungsberichte der Österreichischen Akademie der Wissenschaften*, Phil.-hist. Kl. 364 (1980), pp. 91–108, here pp. 105ff.

174. Nicholas Copernicus, *De revolutionibus*, 'Praefatio', ed. Heribert M. Nobis and Bernhard Sticker (1984), pp. 4ff: calendar reform and universal machine; III, 11, pp. 213ff: computation of times and history. See Hans Blumenberg, *Die Genesis der kopernikanischen Welt*, 2 (1981), pp. 503–606, on the concept of time (the calendar question is mentioned on p. 533, but receives insufficient attention). On the devaluing of the astrolabe, see Borst (as n. 20), pp. 102–5.

175. On the farmers' calendars, see Nils Lithberg, *Computus, med särskild hänsyn till runstaven och den borgerliga kalendern* (1953), pp. 104–210, 244–82; Ludwig Rohner, *Kalendergeschichten und Kalender* (1978), pp. 33–5. In Austria and Switzerland such farmers' calendars even today continue to appear in print.

176. Paul Lehmann, 'Einteilung und Datierung nach Jahrhunderten', in: Lehmann (as n. 42), vol. 1 (1959), pp. 114–29; a more accurate account is given in Johannes Burkhardt, *Die Entstehung der modernen Jahrhundertrechnung. Ursprung und Ausbildung einer historiographischen Technik von Flacius bis Ranke* (1971), pp. 11–28.

177. Friedrich K. Ginzel, *Handbuch der mathematischen und technischen*

Chronologie, 3 (1914), pp. 257–66: the details. The most recent full account is given in: *Gregorian Reform of the Calendar*, ed. George V. Coyne (1983). The latest mathematical and chronological critique is provided by Zemanek (as n. 8), pp. 29–34. The most recent historical survey is provided by Gerhard Römer, 'Kalenderreform und Kalenderstreit im 16. und 17. Jahrhundert', in: *Kalender im Wandel der Zeiten*, ed. Landesbibliothek Baden (1982), pp. 70–84. An original account of the *Missale Romanum* is given in Joachim Mayr, 'Der Computus ecclesiasticus', *Zeitschrift für katholische Theologie*, 77 (1955), pp. 301–30.

178. Giordano Bruno, 'L'asino cillenico del Nolano', in: *Dialoghi italiani*, ed. Giovanni Aquilecchia (1958), p. 922. See Frances A. Yates, *Collected Essays*, 1 (1982); Gerhart von Graevenitz, *Mythos, Zur Geschichte einer Denkgewohnheit* (1987), pp. 1–33.

179. Montaigne, *Essais*, III, 10, ed. Maurice Rat, 2 (1962), pp. 455ff: another person; III, 11, pp. 472ff: time-reckoning. See Hans Blumenberg, *Lebenszeit und Weltzeit* (1986), pp. 148–52. On the later content of *computiste*, see below, n. 193.

180. Joseph Justus Scaliger, *Opus novum de emendatione temporum* (1583), pp. 294–379: ten *computi*; pp. 380–431: his own account, here p. 405, *doctrina annalis*. See Walter E. van Wijk, *Het eerste leerboek der technische tijdrekenkunde* (1954), pp. 1–6.

181. Joseph Justus Scaliger, *Thesaurus temporum*, facsimile edn, ed. Hellmut Rosenfeld and Otto Zeller, 1 (1968), part 1, pp. 1–197: Eusebius and Jerome; vol. 2 (1968), part 3, p. 240: *computus manualis*; p. 117: time; p. 276: *chronologi*; p. 308: Alfonso and Copernicus; pp. 276–309: *tempus historicum*; pp. 273ff, 309ff: *tempus prolepticon*; pp. 274ff: *computus ecclesiasticus*; p. 277: *computatores*. See Anthony T. Grafton, *Joseph Scaliger (1540–1609) and the Humanism of the Later Renaissance*, Diss. phil., Chicago (1975), pp. 173–220 on the historical context; Zemanek (as n. 8), pp. 61–74, 122–9 on the mathematical and chronological consequences up to the present day. Scaliger would number our exemplary date–2 March 1988, 6 p.m. – as 2447223.75.

182. Adalbert Klempt, *Die Säkularisierung der universalhistorischen Auffassung. Zum Wandel des Geschichtsdenkens im 16. und 17. Jahrhundert* (1960), pp. 81–9, in which the search for predecessors overlooks the most important, Bede (see above, n. 63). Kaletsch (as n. 97), p. 80, ignores all predecessors.

183. Bruno von Freytag Löringhoff, 'Wilhelm Schickard und seine Rechenmaschine von 1623', in *Dreihundertfünfzig Jahre Rechenmaschinen*, ed. Martin Graef (1973), pp. 11–20, here pp. 11ff: the quotations. See Michael R. Williams, *A History of Computing Technology* (1985), pp. 123–8.

184. Pascal, 'La Machine arithmétique', in *Oeuvres complètes*, ed. Louis Lafuma (1963), pp. 187–91: the description; 'De l'esprit géométrique', pp. 349–51: time and numbers; *Pensées*, no. 199–72, pp. 526ff: God and man; no. 456–618, p. 561: Jews; no. 821–252, p. 604: *automate*; no. 741–340, p. 596: thinking machine. See also Williams (as n. 183), pp. 128–34; Herbert Heckmann, *Die andere Schöpfung. Geschichte der frühen Automaten in Wirklichkeit und Dichtung* (1982), pp. 90ff. For a diverse,

yet informative, collection of details on the mechanization of the seventeenth-century world-view, see ibid., pp. 165–209.

185. Ludolf von Mackensen, 'Von Pascal zu Hahn. Die Entwicklung von Rechenmaschinen im 17. und 18. Jahrhundert', in: *Dreihundertfünfzig Jahre* (as n. 183), pp. 21–33; Williams (as n. 183), pp. 134–50: generally; pp. 92–7: Schott.

186. *Des Abenteurlichen Simplicissimi Ewig-währender Calender*, facsimile, ed. Klaus Habermann (1967), pp. 4ff: content of the six columns; p. 60 (II)b, 18 March; p. 45a: calendar-makers; p. 11a: definition of time; pp. 29a–31a: date of Creation; p. 39a: Easter; pp. 47a–49a: history of the calendar; p. 91a: lies by the calendar-makers. For a study of the term, see Grimm (as n. 62), vol. 11 (1873), col. 63. On the work, see Habermann, *Beiheft* (supplement to the facsimile edition, 1967), pp. 15–46; Rohner (as n. 175), pp. 119–58. In Johann Peter Hebel, *Der Rheinländische Hausfreund 1808–1819*, facsimile, ed. Ludwig Rohner (1981), p. 146, the phrase *Wir Sternseher und Calendermacher* ('we stargazers and calendar-makers') was intended to be at once more ironical and more scholarly.

187. 'Pseudodoxia epidemica', VI, 1, in: *The Works of Sir Thomas Browne*, ed. Geoffrey Keynes, 2 (1964), p. 409: *exact compute*; p. 403: Bede and Scaliger; VI, 4, p. 419: *computers*; VI, 8, pp. 454ff: *computists*. See Arno Borst, *Der Turmbau von Babel. Geschichte der Meinungen über Ursprung und Vielfalt der Sprachen und Völker*, 3/1 (1960), pp. 1317ff.

188. Swift, 'A Tale of a Tub', ch. 7, in: *Prose Works*, ed. Herbert Davis, 1 (1965), pp. 91–3. More recent references to *computer* are given in: *The Oxford English Dictionary*, ed. James A. Murray et al., 2 (1933), p. 750.

189. Swift, *Gulliver's Travels*, III, 5, ed. Davis (as n. 188), vol. 11 (1965), pp. 182–5, with a comical drawing. See Klaus Arnold, *Geschichtswissenschaft und elektronische Datenverarbeitung, Historische Zeitschrift*, supplement NF 3 (1974), pp. 98–148, here 101ff; as to the context of the enthusiasm for, and criticism of, machines in the eighteenth century, see Heckmann (as n. 184), pp. 235–80.

190. Koselleck (as n. 33), pp. 9–13: on the general change; however, I would assign innovations in historical chronology and in physical and astronomical chronometry to *historical times*, rather than to the putatively *one, natural time*. On the chronometers, see Landes (as n. 5), pp. 129ff, 145–86.

191. Leibniz, *Nouveaux essais sur l'entendement humain*, Preface, in *Die philosophischen Schriften*, ed. Carl I. Gerhardt, 5 (1882) p. 48: the present; II, 14, pp. 138–40: time; II, 16, p. 143: number. See Böhme (as n. 12), pp. 195–256, here p. 199: the Newton quotation; Manfred Eigen, 'Evolution und Zeitlichkeit', in: *Die Zeit* (as n. 24), pp. 35–57; on historical time, see finally Waldemar Voisé, 'On Historical Time in the Works of Leibniz', in: *The Study of Time*, ed. Julius T. Fraser and Nathaniel Lawrence, 2 (1975), pp. 114–21.

192. Giambattista Vico, *La scienza nuova seconda*, I, 1, ed. Fausto Nicolini (1953), pp. 37–72: chronological table; X, 1–2, pp. 357–64: poetic chronology. See Friedrich Meinecke, 'Die Entstehung des Historismus', in: *Werke*, 3, ed. Carl Hinrichs (1965), pp. 53–69, here as in the following without regard to chronological aspects; for a sharper analysis, see

Graevenitz (as n. 178), pp. 65–84; Hans Robert Jauss, 'Mythen des Anfangs: Eine geheime Sehnsucht der Aufklärung', in: Jauss, *Studien zum Epochenwandel der ästhetischen Moderne* (1989), pp. 23–66, here pp. 23–31.

193. Voltaire, 'Essai sur les moeurs et l'esprit des nations', ch. 1, ed. René Pomeau, 1 (1963), pp. 205–9. See Meinecke (as n. 192), pp. 73–115, here p. 76: bourgeoisie. According to Zemanek (as n. 8), p. 93, the Chinese cycle begins somewhat earlier still, in 2637 BC. *Encyclopédie ou dictionnaire raisonné des sciences, des arts et des métiers*, ed. Denis Diderot, 3 (1753), p. 798, defined *comput* as *calcul*, mainly a chronological one for calendrical calculation; *computiste* was now no more than a papal finance official (see above, n. 148).

194. Johann Gottfried Herder, 'Unterhaltungen und Briefe über die ältesten Urkunden', in: *Sämtliche Werke*, ed. Bernhard Suphan, 6 (1883), pp. 180–7. See Meinecke (as n. 192), pp. 359–86; Graevenitz (as n. 178), pp. 84–8. Other references to the German word 'Zeitrechnung' (time-reckoning) are provided by Grimm (as n. 62), vol. 31 (1956), cols 570ff.

195. Serge Bianchi, *La Révolution culturelle de l'an II. Élites et peuple 1789–1799* (1982), pp. 198–203; Zemanek (as n. 8), pp. 100ff; Michael Meinzer, 'Der französische Revolutionskalender und die "Neue Zeit"', in: *Die Französische Revolution als Bruch des gesellschaftlichen Bewusstseins*, ed. Reinhart Koselleck and Rolf Reichardt (1988), pp. 23–71; Hans Maier, *Die christliche Zeitrechnung* (1991), pp. 45–55, 100–7. The final quotation is from Georges Duby and Guy Lardreau, *Dialogues* (1980).

196. Leopold von Ranke, 'Über die Epochen der neueren Geschichte', in: *Aus Werk und Nachlass*, ed. Theodor Schieder et al., 2 (1971), pp. 58–63: epochs; see Burkhardt (as n. 176), pp. 101–9. Jacob Burckhardt, *Über das Studium der Geschichte*, ed. Peter Ganz (1982), p. 108: instrument; p. 276: clock. The final quotation is from Schieder (as n. 21), p. 80. On Mommsen, see above, n. 61; on Krusch, see nn. 37, 38 and 52. On the analogous reaction of German romantics, see Peter Utz, 'Das Ticken des Textes. Zur literarischen Wahrnehmung der Zeit', *Schweizer Monatshefte*. (1990), pp. 649–62 (reference by Gustav Siebenmann).

197. Henning Eichberg, 'Der Umbruch des Bewegungsverhaltens. Leibesübungen, Spiele und Tänze in der Industriellen Revolution', in: *Verhaltenswandel in der Industriellen Revolution*, ed. August Nitschke (1975), pp. 118–35; Landes (as n. 5), pp. 4–6, 130ff. I follow Maurice (as n. 162), vol. 1, p. 284, in dating the beginnings of the stop-watch later than Eichberg.

198. Thomas Nipperdey, *Deutsche Geschichte 1800–1866. Bürgerwelt und starker Staat* (1985), pp. 227–30, on the beginnings. Rolf Hackstein, *Arbeitswissenschaft im Umriss*, 2 (1977), pp. 412–24: Taylor. For a discussion of the context, see David S. Landes, *The Unbound Prometheus* (1969), pp. 318–22.

199. H. G. Wells, *The Time Machine*, ch. 3, in: The Collected Essex Edition, 16 (1927), p. 21, and ch. 12, p. 94: grandfather clock; ch. 4, p. 32: year of the Lord; ch. 11, p. 88: day clocks. See Michael Salewski, *Zeitgeist und Zeitmaschine. Science Fiction und Geschichte* (1986), pp. 121–42.

200. Rolf Oberliesen, *Information, Daten und Signale. Geschichte technischer Informationsverarbeitung* (1982), pp. 195–202, 212–48; Williams (as n.

183), pp. 150–8: *arithmomètre* of Charles X. Thomas and *Comptometer* by Dorr E. Felt. *Hollerith's 'Counters': The Origins of Digital Computers. Selected Papers*, ed. Brian Randell (1973), pp. 135ff. The quotation from Burke is given in full in Borst (as n. 62), p. 663.

201. *A Supplement to the Oxford English Dictionary*, ed. Robert W. Burchfield, 1 (1972), p. 601: the *computer* of 1897. On the astronomers' suggestion: *Das zweite Vatikanische Konzil. Konstitutionen, Dekrete und Erklärungen*, ed. Herbert Vorgrimler, 1 (1966), pp. 108ff. In 1963, in its Latin comments on the calendar reform (ibid., pp. 106–9) and on the ecclesiastical year (pp. 86–95), the Vaticanum itself no longer used the word *computus*.

202. Oberliesen (as n. 200), p. 219: *Statistical Computer*, p. 228: Hollerith's company names (reference by Lothar Burchardt).

203. Heidegger, *Being and Time* (1967), pp. 468–71. See Charles M. Sherover, *The Human Experience of Time. The Development of Its Philosophical Meaning* (1975), pp. 455–65; for a more critical account, see Ernst Pöppel, 'Erlebte Zeit und die Zeit überhaupt. Ein Versuch der Integration', in: *Die Zeit* (as n. 24), pp. 369–82.

204. *The Origins* (as n. 200), pp. 241–6: *Computer* alongside *Computing Machine* in George R. Stibitz (1940), pp. 305–25: similarly in John V. Atanasoff (1940); but *expert computer* (p. 306) meant a person; pp. 355–364: *Computor* (*sic*) alongside *Computing Device* in John von Neumann (1945).

205. Paul Robert, *Dictionnaire alphabétique et analogique de la langue française*, 6 (1985), p. 967; but see below, n. 210. The German usage was determined for a while by Karl Steinbuch, *Die informierte Gesellschaft. Geschichte und Zukunft der Nachrichtentechnik* (1968), p. 151, with the proposal to avoid the terms *Elektronengehirn* ('electronic brain') and *Denkmaschine* ('thought-machine'), adopt the term *computers* for machines, and reserve the word *Rechner* (reckoners) for people.

206. *Abacus elements* in Atanasoff (1940), in: *The Origins* (as n. 200), p. 308. Goldstine (as n. 8), p. 39, gives a careful account of the abacus as a computer; for an unconcerned account, see Edgar P. Vorndran, *Entwicklungsgeschichte des Computers* (1982), pp. 19–22. On the predominantly non-numerical mode of operation of the more recent computers, see Weizenbaum (as n. 8). On Bede, see above, n. 55; on Gerbert, see n. 92.

207. Lewis Mumford, *The Myth of the Machine* (1967), pp. 265, 286ff, 296, exaggerates the continuity between the mechanical clock and the computer; the same is true of Weizenbaum (as n. 8), pp. 21–7; Peter Gendolla, 'Die Einrichtung der Zeit. Gedanken über ein Prinzip der Räderuhr', in: *Augenblick und Zeitpunkt. Studien zur Zeitstruktur und Zeitmetaphorik in Kunst und Wissenschaften*, ed. Christian W. Thomsen and Hans Holländer (1984), pp. 47–58, here pp. 53ff. The breakthrough is described from a medieval perspective by Seibt (as n. 161), pp. 183–5, and from a more modern perspective by Landes (as n. 5), pp. 186ff, 352ff, 376ff, who makes a 'quartz revolution' from it. The most expert and considered account of the 'corrected second' is given by Zemanek (as n. 8), pp. 103–10.

208. Goldstine (as n. 8), pp. 342–7, with the pride of the pioneer who sought

to distinguish a separate 'computer revolution' from the industrial revolution. A more judicious account is provided by Carlo Schmid, 'Die zweite Industrielle Revolution', in: *Propyläen-Weltgeschichte*, ed. Golo Mann, 10 (1961), pp. 423–52, here pp. 438–44.

209. Herman H. Goldstine, *New and Full Moons, 1001 B.C. to A.D. 1651* (1973), pp. vff: astronomical chronology. On its modern prehistory, see Goldstine (as n. 8), pp. 8, 27–30, 108, 327. On Hermann, see Borst (as n. 98), pp. 436–40. The suggestion regarding historical chronology is given in Carl A. Lückerath, 'Prolegomena zur elektronischen Datenverarbeitung im Bereich der Geschichtswissenschaft', *Historische Zeitschrift*, 207 (1968), pp. 265–96, here pp. 284ff. A passionate plea for Scaliger's Julian days and the French revolutionary calendar is made by the writer Arno Schmidt, *Trommler beim Zaren* (1966), pp. 183–91, 196–206. *Epochenschwelle und Epochenbewusstsein*, ed. Reinhart Herzog and Reinhart Kosellek (*Poetik und Hermeneutik*, 12, 1987) is not interested in Scaliger's questions.

210. Fernand Braudel, 'History and the Social Sciences. The *Longue Durée*', in: Braudel, *On History* (1980), pp. 25–54, here p. 43: the quotation. The original of 1958 used *machine à calculer*, not *ordinateur*, for the computer. For an account of the 'cliometric' consequences, see Michael Erbe, *Zur neueren französischen Sozialgeschichtsforschung. Die Gruppe um die 'Annales'* (1979), pp. 94–106.

211. Weizenbaum (as n. 8), p. 9: the quotations. Ernst Jünger, *An der Zeitmauer* (1959), p. 136, scorns the effects of the *Rechenmaschinen* ('calculating machines'), but on p. 19–71 recognizes one of the numerous counter-movements, the recent increase in astrological tendencies.

212. *Brockhaus Enzyklopädie*, 4 (1987), pp. 651–3. The same unreflected usage is demonstrated by the medievalist Arnold in 1974 (as n. 189), pp. 102ff.

213. Peter-Johannes Schuler, 'Datierung von Urkunden', in: *Lexikon des Mittelalters*, 3 (1986), cols 575–80: date. Esch (as n. 153), pp. 321–32: generation.

214. Zemanek (as n. 8), pp. 11ff, 47, 110–14; the last quotation, p. 114.

215. Julius T. Fraser, *Time the Familiar Stranger* (1987), pp. 323–43.

216. Reinhart Koselleck, 'Wie neu ist die Neuzeit?' *Historische Zeitschrift*, 251 (1990), pp. 539–53.

217. Ingeborg Bachmann, 'Die gestundete Zeit', in: *Werke*, ed. Christine Koschel et al., 1 (1978), p. 37, first published in 1953, alluding to Heidegger's closing sentence (as n. 203), §83, p. 577. Quoted separately by Horst Fuhrmann, *Einladung ins Mittelalter* (1987), p. 22.

Index